# Basic Concepts
# in Relativity

## The Fifth Solvay Physics Conference
### Brussels, Belgium, 23–29 October 1927
#### Sponsored by the Solvay International Institute of Physics.

1. A. Piccard
2. E. Henriot
3. P. Ehrenfest
4. Ed. Herzen
5. Th. De Donder
6. E. Schrödinger
7. E. Verschaffelt
8. W. Pauli

9. W. Heisenberg
10. R. H. Fowler
11. L. Brillouin
12. P. Debye
13. M. Knudsen
14. W. L. Bragg
15. H. A. Kramers
16. P. A. M. Dirac

17. A. H. Compton
18. L. de Broglie
19. M. Born
20. N. Bohr
21. I. Langmuir
22. M. Planck
23. Mme. Curie
24. H. A. Lorentz

25. A. Einstein
26. P. Langevin
27. Ch. E. Guye
28. C. T. R. Wilson
29. O. W. Richardson

Absent: Sir. W. H. Bragg,
H. Deslandres et E. Van Aubel

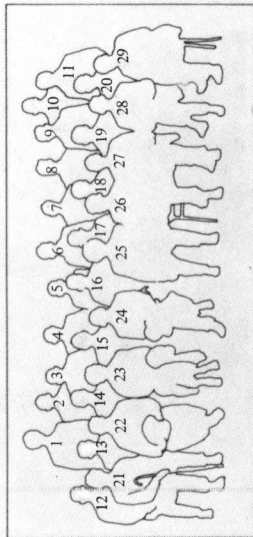

# Basic Concepts in Relativity

**Robert Resnick**
*Rensselaer Polytechnic Institute*

**David Halliday**
*Emeritus*
*University of Pittsburgh*

MACMILLAN PUBLISHING COMPANY
NEW YORK

Maxwell Macmillan Canada
TORONTO

Maxwell Macmillan International
NEW YORK   OXFORD   SINGAPORE   SYDNEY

Macmillan Publishing Company
866 Third Avenue, New York, New York 10022

Maxwell Macmillan Canada, Inc.
1200 Eglinton Avenue, E.
Suite 200
Don Mills, Ontario M3C 3N1

LIBRARY OF CONGRESS CATALOGING IN PUBLICATION DATA

Resnick, Robert (date)
   [Basic concepts in relativity and early quantum mechanics.
  Selections]
   Basic concepts in relativity / Robert Resnick, David Halliday.
     p.    cm.
   "A portion of this book is reprinted from Basic concepts in relativity and early quantum
mechanics, second edition"–T.p. verso.
   Includes index.
   ISBN 0-02-399345-6
   1. Relativity (Physics)   I. Halliday, David (date).
II. Title
QC173.55.R4525   1992
530.1'1 — dc20
                                        91-43563
                                           CIP

Printing:  1  2  3  4  5  6  7  8    Year:  2  3  4  5  6  7  8  9  0  1

# Preface

This text is extracted from Robert Resnick and David Halliday *Basic Concepts in Relativity and Early Quantum Theory,* Second Edition, Revised Printing, Macmillan Publishers (1992). It is meant to meet the need for a flexible supplement to introductory classical physics courses or to beginning quantum physics courses that typically present only a brief exposition of relativity or none at all. There is a growing trend to include relativity in the introductory physics course and to introduce it well before the treatment of modern quantum physics. Usually the material in relativity is spread out throughout the text. This text makes it possible to present a significant treatment of relativity coherently in one place, either in the introductory course or at the beginning of a modern physics course. There are other effective ways to use this text, of course, such as in minicourses and in summer courses for teachers.

In Chapters 1, 2, and 3 we examine the experimental background to relativity, the development of the special theory of relativity, and the experimental confirmation of relativistic predictions. We see that classical mechanics breaks down in the region of very high speeds and that relativistic mechanics is a generalization that includes the classical laws as a special case. Gradually, the student will develop a physical feeling for the principles of relativity. Three supplementary topics are included as well, on spacetime diagrams, on the twin paradox, and on general relativity. These are intended to serve as optional supplementary material for instructors seeking greater depth or breadth of treatment and to satisfy the typical student's interest in these topics.

There is a large, rich, and varied set of questions and problems for discussion and assignment purposes. The more difficult ones have been starred, and a set of problems for hand-held programmable calculators (designated with a C) has been included as well. There is also a large set of worked-out examples in the text, selected to be effective in enhancing the student's understanding and problem-solving ability. The figures have been redone for greater clarity and improved layout.

# Contents

# 3   Relativistic Dynamics                                    104

## Supplementary Topic **A**
### The Geometric Representation of Spacetime                141

## Supplementary Topic **B**
### The Twin Paradox                                          156

## Supplementary Topic **C**
### The Principle of Equivalence and the General Theory of Relativity                                       167

## CHAPTER 1

# The Experimental Background of the Theory of Special Relativity

*. . . whenever energy is transmitted from one body to another in time, there must be a medium or substance in which the energy exists after it leaves one body and before it reaches the other. . . .*

*J. C. Maxwell (1873)*

*. . . I came to the opinion quite some time ago that Fresnel's idea, hypothesizing a motionless ether, is on the right path.*

*H. A. Lorentz (1895)*

*The introduction of a "luminiferous ether" will prove to be superfluous inasmuch as the view here developed will not require an "absolute stationary space". . . .*
*A. Einstein (1905)*

## 1-1 Introduction

To send a signal from one point to another as fast as possible, we use a beam of light or some other electromagnetic radiation such as a radio wave. *No faster method of signaling has ever been discovered.* This experimental fact suggests that the speed of light in free space ($c = 3.00 \times 10^8$ m/s)* is an appropriate limiting reference speed with respect to which other speeds, such as the speeds of particles or of mechanical waves, can be compared.

In the macroscopic world of our ordinary daily experiences, the speed $u$ of moving objects or mechanical waves with respect to any observer is always much less than $c$. For example, a satellite circling the earth may move at 16,000 mi/h with respect to the earth; here $u/c = 0.000024$. Sound waves in air at room temperature travel at 343 m/s through the air, so that $u/c = 0.0000011$. It was in this ever-present, but limited, macroscopic environment that our ideas about space and time were first formulated and in which Newton developed his system of mechanics.

In the microscopic world, however, it is possible to find particles whose speeds are quite close to that of light. For an electron accelerated through a 10-million-volt

*In October 1983, the speed of light was adopted as a defined standard and assigned the value of (exactly) $2.99792458 \times 10^8$ m/s.

potential difference, a value reasonably easy to obtain, we find that $u/c = 0.9988$. We cannot be certain without direct experimental test that Newtonian mechanics can be safely extrapolated from the ordinary region of low speeds ($u/c << 1$) in which it was developed to this high-speed region ($u/c \rightarrow 1$). Experiment shows, in fact, that Newtonian mechanics does *not* predict the correct answers when it is applied to such fast particles. Indeed, in Newtonian mechanics there is no limit in principle to the speed attainable by a particle, so that the speed of light $c$ plays no special role at all. For example, if the energy of the 10-MeV electron above is increased by a factor of four (to 40 MeV), experiment shows that the speed is not doubled to $1.9976c$, as we might expect from the Newtonian relation $K = \frac{1}{2}m_e v^2$, but remains below $c$; it increases only from $0.9988c$ to $0.9999c$, a change of $0.11$ percent. Or, if the 10-MeV electron moves at right angles to a magnetic field of 2.0 T, the measured radius of curvature of its path is not 0.53 cm (as may be computed from the classical relation $r = m_e v/qB$) but, instead, 1.7 cm. Hence, no matter how well Newtonian mechanics may work at low speeds, it fails badly as $u/c \rightarrow 1$.

In 1905 Albert Einstein published his special theory of relativity. Although motivated by a desire to gain deeper insight into the nature of electromagnetism, Einstein, in this theory, extended and generalized Newtonian mechanics as well. He correctly predicted the results of mechanical experiments over the complete range of speeds from $u/c = 0$ to $u/c \rightarrow 1$. Newtonian mechanics was revealed to be an important special case of a more general theory. In developing this theory of relativity, Einstein critically examined the procedures used to measure length and time intervals. These procedures require the use of light signals and, in fact, an assumption about the way light is propagated is one of the two central hypotheses on which the theory is based. His theory resulted in a completely new view of the nature of space and time.

The connection between mechanics and electromagnetism is not surprising because light, which (as we shall see) plays a basic role in making the fundamental space and time measurements that underlie mechanics, is an electromagnetic phenomenon. However, our low-speed Newtonian environment is so much a part of our daily life that almost everyone has some conceptual difficulty in understanding Einstein's ideas of space-time when he or she first studies them. Einstein may have put his finger on the difficulty when he said: "Common sense is that layer of prejudices laid down in the mind prior to the age of eighteen." Indeed, it has been said that every great theory begins as a heresy and ends as a prejudice. The ideas of motion of Galileo and Newton may very well have passed through such a history already. More than three-quarters of a century of experimentation and application has removed special relativity theory from the heresy stage and put it on a sound conceptual and practical basis. Furthermore, we shall show that a careful analysis of the basic assumptions of Einstein and of Newton makes it clear that the assumptions of Einstein are really much more reasonable than those of Newton.

In the following pages, we shall develop the experimental basis for the ideas of special relativity theory. Because, in retrospect, we found that Newtonian mechanics fails when applied to high-speed particles, it seems wise to begin by examining the foundations of Newtonian mechanics. Perhaps, in this way, we can find clues as to how it might be generalized to yield correct results at high speeds while still maintaining its excellent agreement with experiment at low speeds.

# 1-2 Galilean Transformations

Let us begin by considering a physical *event*. An event is something that happens independent of the reference frame we might use to describe it. For concreteness, we can imagine the event to be a collision of two particles or the flashing of a tiny light source. The event happens at a point in space and at an instant in time. We specify an event by four (space-time) measurements in a particular frame of reference, say, the position numbers $x, y, z$ and the time $t$. For example, the collision of two particles may occur at $x = 1$ m, $y = 4$ m, $z = 11$ m, and at time $t = 7$ s in one frame of reference (for example, a laboratory on earth), so that the four numbers (1, 4, 11, 7) specify the event in that reference frame. The same event observed from a different reference frame (for example, an airplane flying overhead) would also be specified by four numbers, although the numbers might be different than those in the laboratory frame. Thus, if we are to describe events, our first step is to establish a frame of reference.

We define an *inertial system* as a frame of reference in which the law of inertia— Newton's first law—holds. In such a system, which we may also describe as an *unaccelerated* system, a body that is acted on by zero net external force will move with a constant velocity. Newton assumed that a frame of reference fixed with respect to the stars is an inertial system. A rocketship drifting in outer space, without spinning and with its engines cut off, provides an ideal inertial system. Any frame moving at constant velocity with respect to such a system is also an inertial frame. However, frames accelerating with respect to such a system are *not* inertial.

In practice, we can often neglect the small (acceleration) effects due to the rotation and the orbital motion of the earth and to solar motion. Thus, we may regard any set of axes fixed on the earth as forming (approximately) an inertial coordinate system. Likewise, any set of axes moving at uniform velocity with respect to the earth, as in a train, ship, or airplane, will be (nearly) inertial, because motion at uniform velocity does not introduce acceleration. However, a system of axes that accelerates with respect to the earth, such as one fixed to a spinning merry-go-round or to an accelerating car, is *not* an inertial system. A particle acted on by zero net external force will not move in a straight line with constant speed according to an observer in such a noninertial system.

The special theory of relativity, which we consider here, deals only with the description of events by observers in inertial reference frames. The objects whose motions we study may be accelerating with respect to such frames, but the frames themselves are unaccelerated. The general theory of relativity, presented by Einstein in 1917, concerns itself with all frames of reference, including noninertial ones. See Supplementary Topic C for a brief discussion of general relativity theory.

Consider now an inertial frame $S$ and another inertial frame $S'$ that moves at a constant velocity $\mathbf{v}$ with respect to $S$, as shown in Fig. 1-1. For convenience, we choose the three sets of axes to be parallel and allow their relative motion to be

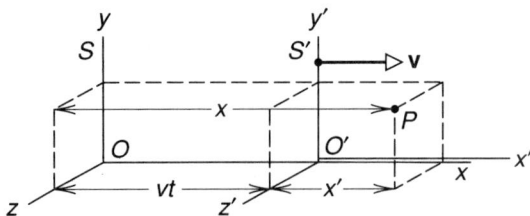

**FIGURE 1-1.** Two inertial frames with a common x-x' axis and with the y-y' and z-z' axes parallel. As seen from frame S, frame S' is moving in the positive x direction at speed v. Similarly, as seen from frame S', frame S is moving in the negative x' direction at this same speed. Point P suggests an *event*, whose space-time coordinates may be measured by each observer. The origins O and O' coincide at time t = 0, t' = 0.

along the common x-x' axis. We can easily generalize to arbitrary orientations and relative velocity of the frames later, but the physical principles involved are not affected by the particular simple choice we make at present. Note also that we can just as well regard S as moving with velocity −**v** with respect to S' as we can regard S' as moving with velocity **v** with respect to S.

Let an event occur at point P, whose space and time coordinates are measured in each inertial frame. An observer attached to S specifies by means of meter sticks and clocks, for instance, the location and time of occurrence of this event, ascribing space coordinates x, y, and z and time t to it. An observer attached to S', using his or her measuring instruments, specifies the *same* event by space-time coordinates x', y', z', and t'. The coordinates x, y, z will give the position of P relative to the origin O as measured by observer S, and t will be the time of occurrence of P that observer S records with his or her clocks. The coordinates x', y', and z' likewise refer the position of P to the origin O' and the time of P, t', to the clocks of inertial observer S'.

We now ask what the relationship is between the measurements x, y, z, t and x', y', z', t'. The two inertial observers use meter sticks, which have been compared and calibrated against one another, and clocks, which have been synchronized and calibrated against one another. The classical procedure, which we look at more critically later, is to assume thereby that length intervals and time intervals are absolute, that is, that they are the same for all inertial observers of the same events. For example, if meter sticks are of the same length when compared at rest with respect to one another, it is implicitly assumed that they are of the same length when compared in relative motion to one another. Similarly, if clocks are calibrated and synchronized when at rest, it is assumed that their readings and rates will agree thereafter, even if they are put in relative motion with respect to one another. These are examples of the "common sense" assumptions of classical theory.

We can state these results explicitly, as follows. For simplicity, let us say that the clocks of each observer read zero at the instant that the origins O and O' of the frames S and S', which are in relative motion, coincide. Then the *Galilean coordinate transformations*, which relate the measurements x, y, z, t to x', y', z', t', are

$$x' = x - vt$$
$$y' = y \qquad\qquad (1\text{-}1a)$$
$$z' = z.$$

These equations agree with our classical intuition, the basis of which is easily seen from Fig. 1-1. It is assumed that time can be defined independently of any particular frame of reference. This is an implicit assumption of classical physics, which is expressed in the transformation equations by the absence of a transformation for $t$. We can make this assumption of the universal nature of time explicit by adding to the Galilean transformations the equation

$$t' = t. \qquad\qquad (1\text{-}1b)$$

It follows at once from Eqs. 1-1a and 1-1b that the time interval between occurrence of two given events, say, $P$ and $Q$, is the same for each observer, that is,

$$t'_P - t'_Q = t_P - t_Q, \qquad\qquad (1\text{-}2a)$$

and that the distance, or space interval, between two points, say, $A$ and $B$, measured at a given instant, is the same for each observer, that is,

$$x'_B - x'_A = x_B - x_A. \qquad\qquad (1\text{-}2b)$$

We have assumed for simplicity that points $A$ and $B$ both lie on the $x$ axis, or on a line parallel to that axis.

## EXAMPLE 1.

*A Galilean Result.* Derive the classical space interval result, Eq. 1-2b.

Let $A$ and $B$ be the end points of a rod, for example, which is at rest in the $S$ frame, parallel to the common $x$-$x'$ axis. Then, the primed observer, for whom the rod is moving with velocity $-\mathbf{v}$, will measure the endpoint locations as $x'_B$ and $x'_A$, whereas the unprimed observer locates them at $x_B$ and $x_A$. Using the Galilean transformations, however, we find that

$$x'_B = x_B - vt_B \quad \text{and} \quad x'_A = x_A - vt_A,$$

so that

$$x'_B - x'_A = x_B - x_A - v(t_B - t_A).$$

Since the two end points, $A$ and $B$, are measured at the same instant, we must put $t_A = t_B$, and we obtain

$$x'_B - x'_A = x_B - x_A,$$

as found above.

Or, we can imagine the rod to be at rest in the primed frame, and moving therefore with velocity $\mathbf{v}$ with respect to the unprimed observer. Then the Galilean transformations, which can be written equivalently as

$$x = x' + vt'$$
$$y = y' \qquad\qquad (1\text{-}3)$$
$$z = z'$$
$$t = t',$$

give us $x_B = x'_B + vt'_B$ and $x_A = x'_A + vt'_A$ and, with $t'_A = t'_B$, we once again obtain $x_B - x_A = x'_B - x'_A$.

Notice carefully that two measurements (the end points $x'_A$, $x'_B$ or $x_A$, $x_B$) are made for each observer and that we assumed they were made at the same time $(t_A = t_B$, or $t'_A = t'_B)$. The assumption that the measurements are made at the same time—that is, simultaneously—is a crucial part of our definition of the length of the moving rod. Surely we should not measure the locations of the end points at different times to get the length of the moving rod; it would be like measuring the location of the tail of a swimming fish at one instant and of its head at another instant in order to determine its length (see Fig. 1-2).

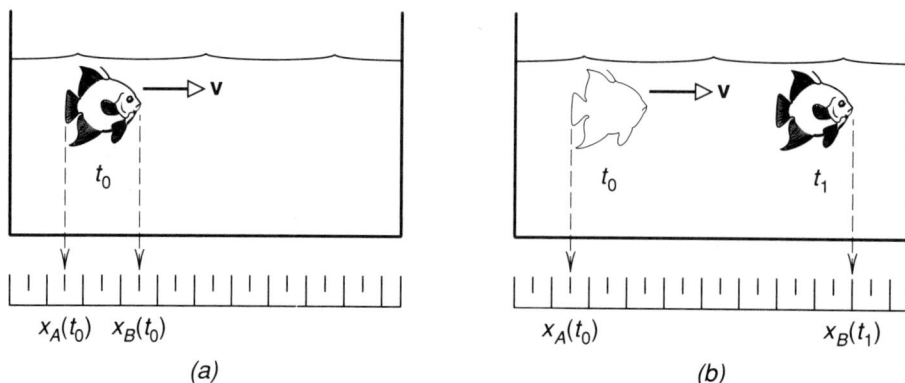

**FIGURE 1-2.**   To measure the length of a swimming fish, one must mark the positions of its head and tail simultaneously (*a*), rather than at arbitrary times (*b*).

The time-interval and space-interval measurements made above are absolutes according to the Galilean transformation; that is, they are the same for all inertial observers, the relative velocity **v** of the frames being arbitrary and not entering into the results. When we add to this result the assumption of classical physics that the mass of a body is a constant, independent of its motion with respect to an observer, then we can conclude that classical mechanics and the Galilean transformations imply that length, mass, and time—the three basic quantities in mechanics—are all independent of the relative motion of the measurer (or observer).

## 1-3 Newtonian Relativity

How do the measurements of different inertial observers compare with regard to velocities and accelerations of objects? The position of a particle in motion is a function of time, so we can express particle velocity and acceleration in terms of time derivatives of position. We need only carry out successive time differentiations of the Galilean transformations. The velocity transformation follows at once. Starting from

$$x' = x - vt,$$

differentiation with respect to $t$ gives

$$\frac{dx'}{dt} = \frac{dx}{dt} - v.$$

But, because $t = t'$, the operation $d/dt$ is identical to the operation $d/dt'$, so

$$\frac{dx'}{dt} = \frac{dx'}{dt'}.$$

Therefore,

$$\frac{dx'}{dt'} = \frac{dx}{dt} - v.$$

Similarly,

$$\frac{dy'}{dt'} = \frac{dy}{dt}$$

and

$$\frac{dz'}{dt'} = \frac{dz}{dt}.$$

However, $dx'/dt' = u'_x$, the $x$ component of the velocity measured in $S'$, and $dx/dt = u_x$, the $x$ component of the velocity measured in $S$, and so on, so that we have simply the *classical velocity addition theorem*:

$$u'_x = u_x - v$$
$$u'_y = u_y \qquad\qquad (1\text{-}4)$$
$$u'_z = u_z.$$

Clearly, in the more general case in which **v,** the relative velocity of the frames, has components along all three axes, we would obtain the more general (vector) result

$$\mathbf{u'} = \mathbf{u} - \mathbf{v}. \qquad\qquad (1\text{-}5)$$

You have already encountered many examples of this. (See *Physics*, Part I Sec. 4-6.*) Thus, the velocity of an airplane with respect to the air (**u'**) equals the velocity of the plane with respect to the ground (**u**) minus the velocity of the air with respect to the ground (**v**).

## EXAMPLE 2.

*Adding Velocities Classically (I).* A passenger walks forward along the aisle of a train at a speed of 3.5 km/h as the train moves along a straight track at a constant speed of 92.5 km/h with respect to the ground. What is the passenger's speed with respect to the ground?

Let us choose the train to be the primed frame so that $u'_x = 3.5$ km/h. The primed frame moves forward with respect to the ground (unprimed frame) at a speed $v = 92.5$ km/h. Hence, the passenger's speed with respect to ground is

$$u_x = u'_x + v = 3.5 \text{ km/h} + 92.5 \text{ km/h}$$
$$= 96.0 \text{ km/h}.$$

## EXAMPLE 3.

*Adding Velocities Classically (II).* Two electrons are ejected in opposite directions from radioactive atoms in a sample of radioactive material at rest in the laboratory. Each electron has a speed of $0.67c$ as measured by a laboratory observer. What is the speed of one electron as measured from the other, according to the classical velocity addition theorem?

Here, we may regard one electron as the $S$ frame, the laboratory as the $S'$ frame, and the other electron as the object whose speed in the $S$ frame is sought (see Fig. 1-3). In the $S'$ frame, the other electron's speed is $0.67c$, moving in the positive $x'$ direction, say, and the speed of the $S$ frame (one electron) is $0.67c$, moving in the negative $x'$ direction. Thus, $u'_x = +0.67c$ and $v = +0.67c$, so that the other electron's speed with respect to the $S$ frame is

$$u_x = u'_x + v = +0.67c + 0.67c$$
$$= +1.34c,$$

according to the classical velocity addition theorem.

*We shall give occasional review references, in this format, to the authors' introductory physics text; see Ref. 1 at the end of the chapter. Other references will be given in the form [3], [4], and so forth.

*(a)*

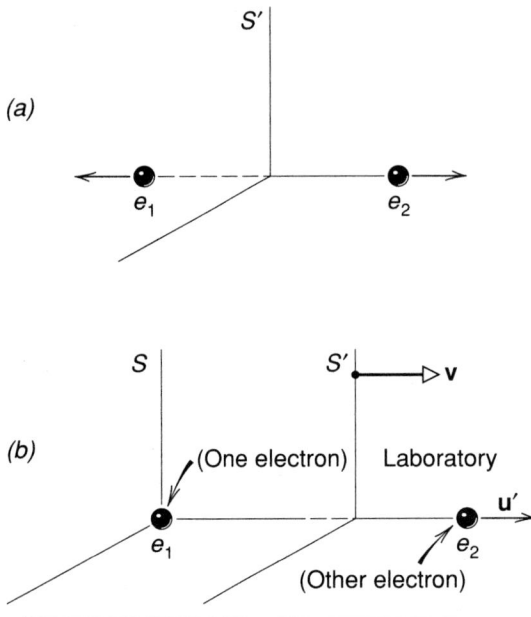

**FIGURE 1-3.**   *(a)* In the laboratory frame, the electrons are observed to move in opposite directions at the same speed. *(b)* In the rest frame, *S*, of one electron, the laboratory moves at a velocity **v**. In the laboratory frame, *S'*, the second electron has a velocity denoted by **u'**. What is the velocity of this second electron as seen by the first?

*(b)*

To obtain the acceleration transformation we merely differentiate the velocity relations (Eq. 1-4). Proceeding as before, we obtain

$$\frac{d}{dt}(u'_x) = \frac{d}{dt}(u_x - v),$$

or, recalling again that $t = t'$,

$$\frac{du'_x}{dt'} = \frac{du_x}{dt}, \qquad v \text{ being a constant,}$$

$$\frac{du'_y}{dt'} = \frac{du_y}{dt},$$

and $\qquad \dfrac{du'_z}{dt'} = \dfrac{du_z}{dt}.$

That is, $a'_x = a_x$, $a'_y = a_y$, and $a'_z = a_z$. Hence, $\mathbf{a'} = \mathbf{a}$. The measured components acceleration of a particle are unaffected by the uniform relative velocity of the reference frames. The same result follows directly from two successive differentiations of Eqs. 1-1*a* and applies generally when **v** has an arbitrary direction, as long as **v** = constant.

We have seen that different velocities are assigned to a particle by different observers when the observers are in relative motion. These velocities always differ by the relative velocity of the two observers, which in the case of inertial observers is a constant velocity. It follows then that when the particle velocity changes, the *change* will be the same for both observers. Thus, they each measure the same *acceleration* for the particle. The acceleration of a particle is the same in *all* reference frames that move relative to one another with constant velocity; that is,

$$\mathbf{a'} = \mathbf{a}. \tag{1-6}$$

In classical physics the *mass* is also unaffected by the motion of the reference frame. Hence, the product $ma$ will be the same for all inertial observers. If $\mathbf{F} = ma$ is taken as the definition of force, then obviously each observer obtains the same measure for each force. With $\mathbf{F} = ma$ and $\mathbf{F}' = ma'$, it follows from Eq. 1-6 that $\mathbf{F} = \mathbf{F}'$. *Newton's laws of motion and the equations of motion of a particle would be exactly the same in all inertial systems.* Since, in classical mechanics, the conservation principles—such as those for energy, linear momentum, and angular momentum—all can be shown to be consequences of Newton's laws, it follows that *the laws of mechanics are the same in all inertial frames.* In more formal language, we say that the laws of mechanics are *invariant* under a Galilean transformation. Let us make sure that we understand just what this paragraph says before we draw some important conclusions from it.

Although different inertial observers will record different velocities for the same particle, and hence different momenta and kinetic energies, they will agree that momentum is conserved in a collision or is not conserved, that mechanical energy is conserved or is not conserved, and so forth. A tennis ball on the court of a moving ocean liner will have a different velocity to a passenger than it has to an observer on shore, and the billiard balls on the table in a home will have different velocities to the player than they have to an observer on a passing train. But, whatever the values of the particle's or system's momentum or mechanical energy may be, when one observer finds that they do not change in an interaction, the other observer will find the same thing. Although the numbers assigned to such things as velocity, momentum, and kinetic energy may be different for different inertial observers, the laws of mechanics (for example, Newton's laws and the conservation principles) will be the same in all inertial systems. This is illustrated in the following example.

## EXAMPLE 4.

*A Collision, Viewed by Two Observers.* A particle of mass $m_1 = 3$ kg, moving at a velocity of $u_1 = +4$ m/s along the $x$ axis of frame $S$, approaches a second particle of mass $m_2 = 1$ kg, moving at a velocity $u_2 = -3$ m/s along this axis. After a head-on collision, we find that $m_2$ has a velocity $U_2 = +3$ m/s along the $x$ axis.

(a) Calculate the expected velocity $U_1$ of $m_1$, after the collision.

We use the law of conservation of momentum. Before the collision the momentum of the system of two particles is

$$P = m_1 u_1 + m_2 u_2$$
$$= (3 \text{ kg})(+4 \text{ m/s}) + (1 \text{ kg})(-3 \text{ m/s})$$
$$= +9 \text{ kg} \cdot \text{m/s}.$$

After the collision the momentum of the system,

$$P = m_1 U_1 + m_2 U_2,$$

is also $+9$ kg $\cdot$ m/s, so that

$$+9 \text{ kg} \cdot \text{m/s} = (3 \text{ kg})(U_1) + (1 \text{ kg})(+3 \text{ m/s})$$

or

$$U_1 = +2 \text{ m/s}, \quad \text{along the } x \text{ axis}.$$

(b) Discuss the collision as seen by observer $S'$ who has a velocity $\mathbf{v}$ of $+2$ m/s relative to $S$ along the $x$ axis.

The four velocities measured by $S'$ can be calculated from the Galilean velocity transformation equation (Eq. 1-5), $\mathbf{u}' = \mathbf{u} - \mathbf{v}$, from which we get

$$u'_1 = u_1 - v = +4 \text{ m/s} - 2 \text{ m/s} = 2 \text{ m/s},$$
$$u'_2 = u_2 - v = -3 \text{ m/s} - 2 \text{ m/s} = -5 \text{m/s},$$
$$U'_1 = U_1 - v = +2 \text{ m/s} - 2 \text{ m/s} = 0,$$
$$U'_2 = U_2 - v = +3 \text{ m/s} - 2 \text{ m/s} = 1 \text{ m/s}.$$

The system momentum in $S'$ is

$$P' = m_1 u'_1 + m_2 u'_2$$
$$= (3 \text{ kg})(2 \text{ m/s}) + (1 \text{ kg})(-5 \text{ m/s})$$
$$= +1 \text{ kg} \cdot \text{m/s}$$

before the collision, and

$$P' = m_1 U'_1 + m_2 U'_2$$

$$= (3 \text{ kg})(0) + (1 \text{ kg})(1 \text{ m/s})$$
$$= +1 \text{ kg} \cdot \text{m/s}$$

after the collision.

Hence, although the velocities and momenta have different numerical values in the two frames, $S$ and $S'$, when momentum is conserved in $S$ it is also conserved in $S'$.

---

An important consequence of the above discussion is that *no mechanical experiments carried out entirely in one inertial frame can tell the observer what the motion of that frame is with respect to any other inertial frame.* The billiard player in a closed boxcar of a train moving uniformly along a straight track cannot tell from the behavior of the balls what the motion of the train is with respect to ground. The tennis player in an enclosed court on an ocean liner moving with uniform velocity (in a calm sea) cannot tell from her game what the motion of the boat is with respect to the water. No matter what the relative motion may be (perhaps none), so long as it is constant, the results will be identical. Of course, we *can* tell what the *relative* velocity of two frames may be by comparing the data the different observers take on the very same event—but then we have not deduced the relative velocity from observations *confined to a single frame.*

Furthermore, there is no way at all of determining the *absolute* velocity of an inertial reference frame from our mechanical experiments. No inertial frame is preferred over any other, for the laws of mechanics are the same in all. Hence, there is no physically definable absolute rest frame. We say that all inertial frames are equivalent as far as mechanics is concerned. The person riding the train cannot tell absolutely whether he alone is moving, or the earth alone is moving past him, or if some combination of motions is involved. Indeed, would you say that you on earth are at rest, that you are moving 30 km/s (the speed of the earth in its orbit about the sun), or that your speed is much greater still (for instance, the sun's speed in its orbit about the galactic center)? Actually, no mechanical experiment can be performed that will detect an absolute velocity through empty space. This result, that we can speak only of the *relative* velocity of one frame with respect to another, and not of an absolute velocity of a frame, is sometimes called *Newtonian relativity.*

Transformation laws, in general, will change many quantities but will leave some others unchanged. These unchanged quantities are called *invariants* of the transformation. In the Galilean transformation laws for the relation between observations made in different inertial frames of reference, for example, acceleration is an invariant and—more important—so are Newton's laws of motion. A statement of what the invariant quantities are is called a relativity principle; it says that for such quantities the reference frames are equivalent to one another, no one having an absolute or privileged status relative to the others. Newton expressed his relativity principle as follows: "The motions of bodies included in a given space are the same amongst themselves, whether that space is at rest or moves uniformly forward in a straight line."

# 1-4 Electromagnetism and Newtonian Relativity

Let us now consider the situation from the electrodynamic point of view. That is, we inquire now whether the laws of physics other than those of mechanics (such as the laws of electromagnetism) are invariant under a Galilean transformation. If so, then the (Newtonian) relativity principle would hold not only for mechanics but for all of physics. That is, no inertial frame would be preferred over any other, and *no type of experiment* in physics, not merely mechanical ones, carried out in a single frame would enable us to determine the velocity of our frame relative to any other frame. There would then be no preferred, or absolute, reference frame.

To see at once that the electromagnetic situation is different from the mechanical one, as far as the Galilean transformations are concerned, consider a pulse of light (that is, an electromagnetic pulse) traveling to the right with respect to the medium through which it is propagated at a speed $c$. A "medium" of light propagation, given the name "ether," was considered necessary, for when the mechanical view of physics dominated physicists' thinking (late nineteenth century and early twentieth century), it was not really accepted that an electromagnetic disturbance could be propagated in empty space. Sound waves, for example, require a medium for propagation. For simplicity, we may regard the "ether" frame, $S$, as an inertial one in which the speed of light $c$ ($= 1/\sqrt{\mu_0 \varepsilon_0}$) has its internationally agreed-upon value; see the footnote on page 1. In a frame $S'$ moving at a constant speed $v$ with respect to this ether frame, an observer would measure a different speed for the light pulse, ranging from $c + v$ to $c - v$, depending on the direction of relative motion, according to the Galilean velocity transformation (Eq. 1-5).

Hence, the speed of light is certainly *not* invariant under a Galilean transformation. If these transformations really do apply to optical or electromagnetic phenomena, then there is one inertial system, and only one, in which the measured speed of light is exactly $c$; that is, there is a unique inertial system in which the so-called ether is at rest. We would then have a physical way of identifying an absolute (or rest) frame and of determining by optical experiments carried out in some other frame what the relative velocity of that frame is with respect to the absolute one.

A more formal way of saying this is as follows. Maxwell's equations of electromagnetism, from which we deduce the electromagnetic wave equation for example, contain the constant $c = 1/\sqrt{\mu_0 \varepsilon_0}$, which is identified as the velocity of propagation of a plane wave in vacuum (see *Physics,* Part II, Sec. 41-8). But such a velocity cannot be the same for observers in different inertial frames, according to the Galilean transformations, so Maxwell's equations and therefore electromagnetic effects will probably not be the same for different inertial observers. But if we accept both the Galilean transformations, and Maxwell's equations as basically correct, then it automatically follows that there exists a unique privileged frame of reference (the "ether" frame) in which Maxwell's equations are valid and in which light is propagated at a speed $c = 1/\sqrt{\mu_0 \varepsilon_0}$.

The situation then seems to be as follows.* The fact that the Galilean relativity principle *does* apply to the Newtonian laws of mechanics but *not* to Maxwell's laws

---

* The treatment here follows closely that of Ref. 2.

of electromagnetism requires us to choose the correct consequences from among the following possibilities.

1. *There is an ether.* A relativity principle exists for mechanics, but *not* for electrodynamics; in electrodynamics there *is* a preferred inertial frame, that is, the ether frame. Should this alternative be correct, the Galilean transformations would apply and we would be able to locate the ether frame experimentally.

2. *Maxwell was wrong.* A relativity principle exists *both* for mechanics and for electrodynamics, but the laws of electrodynamics as given by Maxwell are *not* correct. If this alternative were correct, we ought to be able to perform experiments that show deviations from Maxwell's electrodynamics and reformulate the electromagnetic laws. The Galilean transformations would apply here also.

3. *Newton was wrong.* A relativity principle exists *both* for mechanics and for electrodynamics, but the laws of mechanics as given by Newton are *not* correct. If this alternative is the correct one, we should be able to perform experiments that show deviations from Newtonian mechanics and reformulate the mechanical laws. In that event, the correct transformation laws would not be the Galilean ones (for they are inconsistent with the invariance of Maxwell's equations) but some other ones that are consistent with classical electromagnetism and the new mechanics.

We have already indicated (Section 1-1) that Newtonian mechanics breaks down at high speeds, so you will not be surprised to learn that alternative 3, leading to Einsteinian relativity, is the correct one. In the following sections, we shall look at the experimental bases for rejecting alternatives 1 and 2, as a fruitful prelude to finding the new relativity principle and transformation laws of alternative 3.

## 1-5 Attempts to Locate the Absolute Frame — The Michelson-Morley Experiment

The obvious experiment would be one in which we measure the speed of light in a variety of inertial systems, noting whether the measured speed is different in different systems, and if so, noting especially whether there is evidence for a single unique system — the "ether" frame — in which the speed of light is $c$, the value predicted from electromagnetic theory. A. A. Michelson in 1881 and Michelson and E. W. Morley in 1887 carried out such an experiment [3]. To understand the setting better, let us look a bit further into the "ether" concept.

When we say that the speed of sound in dry air at 0°C is 331.4 m/s, we have in mind an observer, and a corresponding reference system, fixed in the air mass through which the sound wave is moving. The speed of sound for observers moving with respect to this air mass is correctly given by the usual Galilean velocity transformation Eq. 1-5. However, when we say that the speed of light in a vacuum is $2.99792458 \times 10^8$ m/s, it is not at all clear what reference system is implied. A reference system fixed in the medium of propagation of light presents difficulties because, in contrast to sound, no medium seems to exist. However, it seemed inconceivable to nineteenth-century physicists that light and other electromagnetic waves, in contrast to all other kinds of waves, could be propagated without a medium. It seemed to be a logical step to postulate such a medium, called the ether, even though it was necessary to assume unusual properties for it, such as zero density and perfect transparency, to account for its undetectability. The ether was assumed to fill all space and to be the medium with respect to which the speed $c$ applies. It followed then that an observer moving through the ether with velocity $\mathbf{v}$ would measure a velocity $\mathbf{c}'$ for a light beam, where $\mathbf{c}' = \mathbf{c} - \mathbf{v}$. It was this result that the Michelson-Morley experiment was designed to test.

If an ether exists, the spinning and revolving earth should be moving through it. An observer on earth would sense an ''ether wind,'' whose velocity is $\mathbf{v}$ relative to the earth. If we were to assume that $v$ is equal to the earth's orbital speed about the sun, about 30 km/s, then $v/c \cong 10^{-4}$. Optical experiments, which were accurate to the first order in $v/c$, were not able to detect the absolute motion of the earth through the ether, but Fresnel (and later Lorentz) showed how this result could be interpreted in terms of an ether theory. This interpretation had difficulties, however, so that the issue was not really resolved satisfactorily with first-order experiments (experiments whose results depend only on the first power of the ratio $v/c$). It was generally agreed that an unambiguous test of the ether hypothesis would require an experiment that measured the ''second-order'' effect, that is, one that measured $(v/c)^2$. The first-order effect is not large to begin with ($v/c = 10^{-4}$, an effect of one part in 10,000) but the second-order effect is really very small ($v^2/c^2 = 10^{-8}$, an effect of one part in 100 million).

It was A. A. Michelson who invented the optical interferometer whose remarkable sensitivity made such an experiment possible. Michelson first performed the experiment in 1881, and then—in 1887, in collaboration with E. W. Morley—carried out the more precise version of the investigation that was destined to raise troublesome questions for classical physics and that opened the way to a more receptive consideration of relativity theory. For his invention of the interferometer and his many optical experiments, Michelson was awarded the Nobel Prize in Physics in 1907, the first American to be so honored.

Let us now describe the Michelson-Morley experiment. The Michelson interferometer (Fig. 1-4) is fixed on the earth. If we imagine the ''ether'' to be fixed with respect to the sun, then the earth (and interferometer) moves through the ether at a speed of 30 km/s, in different directions in different seasons (Fig. 1-5). For the moment, neglect the earth's spinning motion. The beam of light (plane waves, or parallel rays) from the laboratory source $S$ (fixed with respect to the instrument) is split by the partially silvered mirror $M$ into two coherent beams, beam 1 being transmitted through $M$ and beam 2 being reflected off $M$. Beam 1 is reflected back to $M$ by mirror $M_1$ and beam 2 by mirror $M_2$. Then the returning beam 1 is partially reflected and the returning beam 2 is partially transmitted by $M$ back to a telescope $T$, where they interfere. The interference is constructive or destructive, depending on the phase difference of the beams. The partially silvered mirror surface $M$ is inclined at 45° to the beam directions. If $M_1$ and $M_2$ are very nearly (but not quite) at right angles, we shall observe a fringe system in the telescope (Fig. 1-6) consisting of nearly parallel lines, much as we get from a thin wedge of air between two glass plates.

Let us compute the phase difference between the beams 1 and 2. This difference can arise from two causes, the different path lengths traveled, $l_1$ and $l_2$, and the different speeds of travel with respect to the instrument because of the ''ether wind''

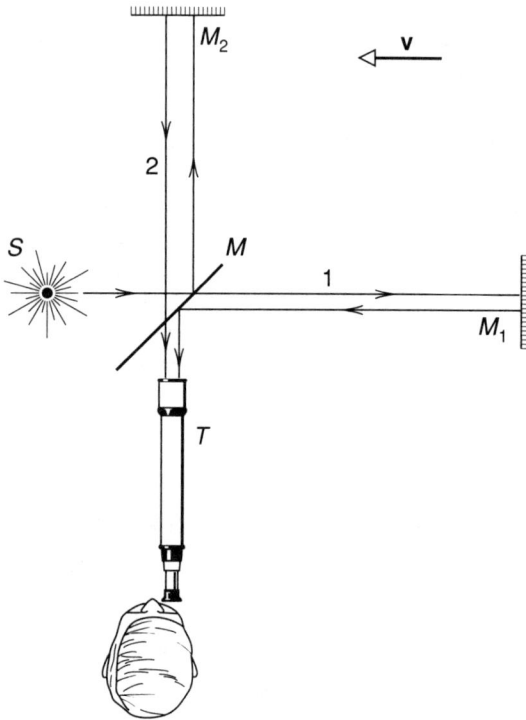

**FIGURE 1-4.**   A simplified version of the Michelson interferometer showing how the beam from the source $S$ is split into two beams by the partially silvered mirror $M$. The beams are reflected by mirrors $M_1$ and $M_2$, returning to the partially silvered mirror. The beams are then transmitted to the telescope $T$, where they interfere, giving rise to a fringe pattern. In this figure, **v** *is the presumed velocity of the ether with respect to the interferometer.*

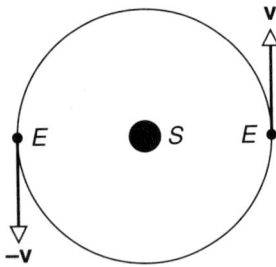

**FIGURE 1-5.**   The earth $E$ moves at an orbital speed of 30 km/s along its nearly circular orbit about the sun $S$, reversing the direction of its velocity every six months.

$v$. The second cause, for the moment, is the crucial one. The different speeds are much like the different cross-stream and up-and-down-stream speeds with respect to shore of a swimmer in a moving stream. The time for beam 1 to travel from $M$ to $M_1$ and back is

$$t_1 = \frac{l_1}{c - v} + \frac{l_1}{c + v} = l_1\left(\frac{2c}{c^2 - v^2}\right) = \frac{2l_1}{c}\left(\frac{1}{1 - v^2/c^2}\right),$$

for the light, whose speed is $c$ in the ether, has an "upstream" speed of $c - v$ with respect to the apparatus and a "downstream" speed of $c + v$. The path of beam 2, traveling from $M$ to $M_2$ and back, is a cross-stream path through the ether, as shown

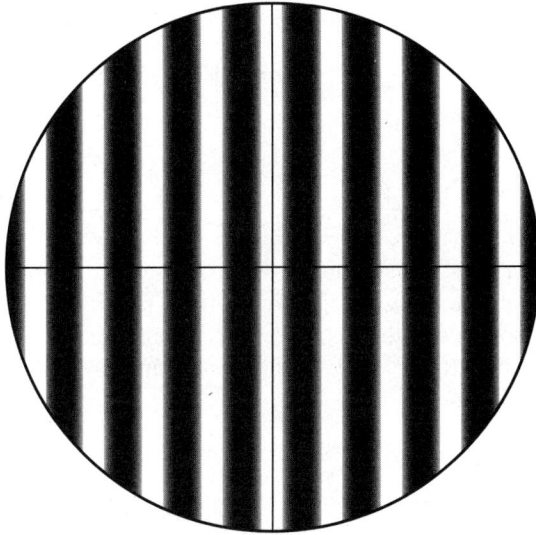

**FIGURE 1-6.**   A typical fringe system seen through the telescope $T$ when $M_1$ and $M_2$ are not quite at right angles.

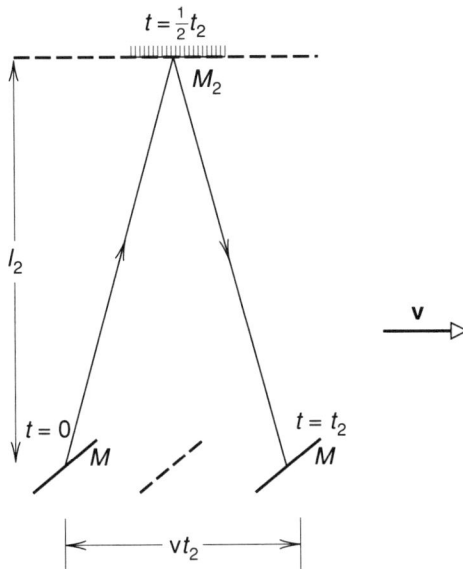

**FIGURE 1-7.**   The cross-stream path of beam 2. The mirrors move through the "ether" at a speed $v$, the light moving through the "ether" at speed $c$. Reflection from the moving mirror automatically gives the cross-stream path. In this figure, **v** *is the presumed velocity of the interferometer with respect to the "ether."*

in Fig. 1-7, enabling the beam to return to the (advancing) mirror $M$. The transit time is given by

$$2\left[l_2^2 + \left(\frac{vt_2}{2}\right)^2\right]^{1/2} = ct_2$$

or

$$t_2 = \frac{2l_2}{\sqrt{c^2 - v^2}} = \frac{2l_2}{c}\frac{1}{\sqrt{1 - v^2/c^2}}.$$

The calculation of $t_2$ is made in the ether frame, that of $t_1$ in the frame of the apparatus. Because time is an absolute in classical physics, this is perfectly acceptable classically. Note that both effects are second-order ones $(v^2/c^2 \cong 10^{-8})$ and are in the same direction (they *increase* the transit time over the case $v = 0$). The difference in transit times is

$$\Delta t = t_2 - t_1 = \frac{2}{c}\left(\frac{l_2}{\sqrt{1 - \beta^2}} - \frac{l_1}{1 - \beta^2}\right) \qquad (1\text{-}7a)$$

in which, for convenience, we have replaced the dimensionless ratio $v/c$ by the symbol $\beta$, often referred to as the *speed parameter*.

Suppose that the instrument is rotated through 90°, thereby making $l_1$ the cross-stream length and $l_2$ the downstream length. If the corresponding times are designated by primes, the same analysis gives the transit-time difference as

$$\Delta t' = t_2' - t_1' = \frac{2}{c}\left(\frac{l_2}{1 - \beta^2} - \frac{l_1}{\sqrt{1 - \beta^2}}\right). \qquad (1\text{-}7b)$$

Hence, *the rotation changes the differences* by

$$\Delta t' - \Delta t = \frac{2}{c}\left(\frac{l_2 + l_1}{1 - \beta^2} - \frac{l_2 + l_1}{\sqrt{1 - \beta^2}}\right).$$

Using the binomial expansion* and dropping terms above the second-order, we find

$$\Delta t' - \Delta t \cong \frac{2}{c}(l_1 + l_2)\left(1 + \beta^2 - 1 - \frac{1}{2}\beta^2\right) = \left(\frac{l_1 + l_2}{c}\right)\beta^2.$$

Therefore, the rotation should cause a shift in the fringe pattern, since it changes the phase relationship between beams 1 and 2.

If the optical path difference between the beams changes by one wavelength, for example, there will be a shift of one fringe across the cross hairs of the viewing telescope. Let $\Delta N$ represent the number of fringes moving past the cross hairs as the pattern shifts. Then, if light of wavelength $\lambda$ is used, so that the period of one vibration is $T = 1/v = \lambda/c$,

$$\Delta N = \frac{\Delta t' - \Delta t}{T} \cong \frac{l_1 + l_2}{cT}\beta^2 = \frac{(l_1 + l_2)\beta^2}{\lambda}. \qquad (1\text{-}8)$$

Michelson and Morley were able to obtain an optical path length, $l_1 + l_2$, of about 22 m. In their experiment the arms were of (nearly) equal length, that is, $l_1 =$

---

* In these cases the binomial expansion (see Appendix, no. 5) gives

$$\frac{1}{1 - \beta^2} = (1 - \beta^2)^{-1} = 1 + \beta^2 + \beta^4 + \beta^6 + \cdots,$$

and

$$\frac{1}{(1 - \beta^2)^{1/2}} = (1 - \beta^2)^{-1/2} = 1 + \frac{1}{2}\beta^2 + \frac{3}{8}\beta^4 + \frac{5}{16}\beta^6 + \cdots.$$

$l_2 = l$, so that $\Delta N = 2l\beta^2/\lambda$. If we choose $\lambda = 5.5 \times 10^{-7}$ m and $\beta(= v/c) = 10^{-4}$, we obtain, from Eq. 1-8,

$$\Delta N = \frac{(22 \text{ m})(10^{-4})^2}{5.5 \times 10^{-7} \text{ m}} = 0.4,$$

or a shift of four-tenths a fringe!

Michelson and Morley mounted the interferometer on a massive stone slab for stability and floated the apparatus in mercury so that it could be rotated smoothly about a central pin. In order to make the light path as long as possible, mirrors were arranged on the slab to reflect the beams back and forth through eight round trips. The fringes were observed under a continuous rotation of the apparatus, and a shift as small as $\frac{1}{100}$ of a fringe definitely could have been detected (see Fig. 1-8).

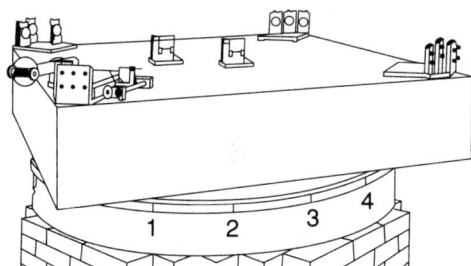

(a)

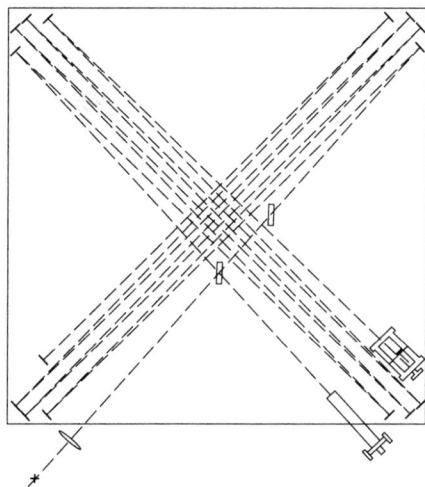

(b)

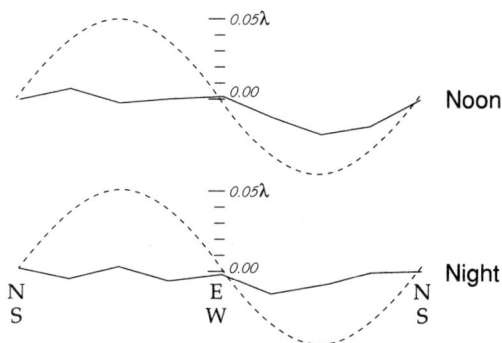

(c)

**FIGURE 1-8.** (a) Mounting of the Michelson-Morley apparatus. (b) Plan view. (c) Observed results. The broken solid lines show the observed fringe shift in the Michelson-Morley experiment as a function of the angle of rotation of the interferometer. The smooth dashed curves—which should be *multiplied by a factor of 8* to bring it to the proper scale—show the fringe shift predicted by the ether hypothesis.

Observations were made day and night (as the earth spins about its axis) and during all seasons of the year (as the earth revolves about the sun), but the expected fringe shift was not observed. Indeed, the experimental conclusion was that *there was no fringe shift at all.*

This null result ($\Delta N = 0$) was such a blow to the ether hypothesis that the experiment was repeated by many workers over a 50-year period. The null result was amply confirmed (see Table 1-1) and provided a great stimulus to theoretical and experimental investigation. In 1958 Cedarholm, Bland, Havens, and Townes [4] carried out an "ether-wind" experiment using microwaves in which they showed that if there is an ether and the earth is moving through it, the earth's speed with respect to the ether would have to be less than $\frac{1}{1000}$ of the earth's orbital speed. This is an improvement of 50 in precision over the best experiment of the Michelson-Morley type. The null result is well established.

Note that the Michelson-Morley experiment depends essentially on the 90° rotation of the interferometer, that is, on interchanging the roles of $l_1$ and $l_2$, as the apparatus moves with a speed $v$ through an "ether." In predicting an expected fringe shift, we took $v$ to be the earth's velocity with respect to an ether fixed with the sun. However, the solar system itself might be in motion with respect to the hypothetical ether. Actually, the experimental results themselves determine the earth's speed with respect to an ether, if indeed there is one, and these results give $v = 0$. Now,

**TABLE 1-1**
Early Trials of the Michelson-Morley Experiment [a]

| Observer | Year | Place | $l$ in Meters | Fringe Shift Predicted by Ether Theory | Upper Limit of Observed Fringe Shift |
|---|---|---|---|---|---|
| Michelson | 1881 | Potsdam | 1.2 | 0.04 | 0.02 |
| Michelson & Morley | 1887 | Cleveland | 11.0 | 0.40 | 0.01 |
| Morley & Miller | 1902–1904 | Cleveland | 32.2 | 1.13 | 0.015 |
| Miller | 1921 | Mt. Wilson | 32.0 | 1.12 | 0.08 |
| Miller | 1923–1924 | Cleveland | 32.0 | 1.12 | 0.030 |
| Miller (sunlight) | 1924 | Cleveland | 32.0 | 1.12 | 0.014 |
| Tomaschek (starlight) | 1924 | Heidelberg | 8.6 | 0.3 | 0.02 |
| Miller | 1925–1926 | Mt. Wilson | 32.0 | 1.12 | 0.088 |
| Kennedy | 1926 | Pasadena and Mt. Wilson | 2.0 | 0.07 | 0.002 |
| Illingworth | 1927 | Pasadena | 2.0 | 0.07 | 0.0004 |
| Piccard & Stahel | 1927 | Mt. Rigi | 2.8 | 0.13 | 0.006 |
| Michelson et al. | 1929 | Mt. Wilson | 25.9 | 0.9 | 0.010 |
| Joos | 1930 | Jena | 21.0 | 0.75 | 0.002 |

[a]From R. S. Shankland, S. W. McCuskey, F. C. Leone, and G. Kuerti, *Rev. Mod. Phys.,* **27,** 167 (1955).

if at some time the velocity were zero in such an ether, no fringe shift would be expected, of course. But the velocity cannot always be zero, since the velocity of the apparatus is *changing* from day to night (as the earth rotates on its axis) and from season to season (as the earth revolves about the sun). Therefore, the experiment does not depend solely on an "absolute" velocity of the earth through an ether, but also depends on the changing velocity of the earth with respect to the "ether." Such a changing motion through the "ether" would be easily detected and measured by the precision experiments, if there were an ether frame. The null result seems to rule out an ether (absolute) frame.

At the end of the last century the ether concept was so deeply ingrained and universally accepted that the null result of the Michelson-Morley experiment came as a real blow. Scientists, anxious to "save the ether," advanced several hypotheses designed to explain this null result without abandoning belief in the ether. We explore these hypotheses in succeeding sections, as a compelling prelude to the need for a totally new approach.

## 1-6 Attempts to "Save the Ether"—the Lorentz-Fitzgerald Contraction

A few years after Michelson and Morley reported the null result of their ether drift experiment, H. A. Lorentz and G. F. Fitzgerald [5] independently proposed a way to account for that result without abandoning the concept of the ether as a preferred reference frame. Although this so-called *Lorentz-Fitzgerald contraction* hypothesis can indeed account for the null Michelson-Morley result, we shall see that it fails to account for the equally crucial null result of a second ether-drift experiment, performed in 1932 by R. J. Kennedy and E. M. Thorndike. We now examine each of these experiments in relationship to the contraction hypothesis.

**The Michelson-Morley Experiment.** The hypothesis advanced by Lorentz and by Fitzgerald was that all bodies are contracted in the direction of motion relative to the stationary ether by a factor $\sqrt{1 - v^2/c^2}$. Again, let the ratio $v/c$ be represented by the symbol $\beta$, so that this factor may be written as $\sqrt{1 - \beta^2}$. Now, if $l^0$ represents the length of a body at rest with respect to the ether (its rest length) and $l$ its length when in motion with respect to the ether, then in the Michelson-Morley experiment

$$l_1 = l_1^0 \sqrt{1 - \beta^2} \quad \text{and} \quad l_2 = l_2^0.$$

This last result follows from the fact that in the hypothesis it was assumed that lengths at right angles to the motion are unaffected by the motion. Then (see Eq. 1-7a)

$$\Delta t = \frac{2}{c} \frac{1}{\sqrt{1 - \beta^2}} (l_2^0 - l_1^0). \tag{1-9a}$$

On 90° rotation we have

$$l_1 = l_1^0 \quad \text{and} \quad l_2 = l_2^0 \sqrt{1 - \beta^2}$$

so that (see Eq. 1-7*b*)

$$\Delta t' = \frac{2}{c} \frac{1}{\sqrt{1 - \beta^2}} (l_2^0 - l_1^0). \tag{1-9b}$$

Hence, no fringe shift should be expected on rotation of the interferometer, for $\Delta t' - \Delta t = 0$.

Lorentz was able to account for such a contraction in terms of his electron theory of matter, but the theory was elaborate and somewhat contrived, and other results predicted by it could not be found experimentally.

**The Kennedy-Thorndike Experiment.** Kennedy and Thorndike carried out an ether-drift experiment using an interferometer with these characteristics:

1. The arms, instead of being (nearly) equal in length as in the Michelson-Morley experiment, were deliberately made as different in length as the coherence of their light source permitted. In their case the length difference was 16 cm.

2. Their interferometer was not free to rotate but was rigidly fixed in the laboratory. The observations consisted of watching for a fringe shift as a function of time as the days and months went by. The instrument had to be sufficiently rigid and stable that its dimensions did not change during these long observation periods.

The theory of the Kennedy-Thorndike experiment is based on the fact that as the earth spins on its axis and orbits the sun, the presumed "ether wind" should change periodically, in both magnitude and direction. Let us suppose that there really is an ether wind and that during a certain interval its magnitude changes from $v$ to $v'$. The predicted fringe shift, accurate to second-order terms in $v/c$ and *taking the Lorentz-Fitzgerald contraction hypothesis fully into account,* is given by

$$\Delta N = \frac{l_1^0 - l_2^0}{\lambda} \left( \frac{v^2 - v'^2}{c^2} \right). \tag{1-10}$$

After monitoring their instrument continuously for intervals ranging from a few days to a month and extending over a period of almost a year, Kennedy and Thorndike concluded that there simply was no observable fringe shift; although the quantity $(v^2 - v'^2)/c^2$ should have changed appreciably during their observation period, the experimental finding was simply that $\Delta N = 0$, in spite of the prediction of Eq. 1-10.

The contraction hypothesis thus fails completely to account for the experimental results. Having failed here, we have no logical basis for invoking the contraction hypothesis to account for the null result of the Michelson-Morley experiment. We must discard this hypothesis completely and account for the null results of both of these important ether-drift experiments from an entirely new point of view, that of the special theory of relativity.

# 1-7 Attempts to "Save the Ether"—the Ether-Drag Hypothesis

Another idea advanced to retain the notion of an ether was that of "ether drag." This hypothesis assumed that the ether frame was attached to all bodies of finite mass, that is, dragged along with such bodies. The assumption of such a "local" ether would automatically give a null result in the Michelson-Morley experiment. Its attraction lay in the fact that it did not require modification of classical mechanics or electromagnetism. However, there were two well-established effects that contradicted the either-drag hypothesis: stellar aberration and the Fizeau convection experiment. Let us consider these effects now, since we must explain them eventually by whatever theory we finally accept.

**Stellar Aberration.** The aberration of light was first reported by the British astronomer James Bradley [6] in 1727. He observed that (with respect to astronomical coordinates fixed with respect to the earth) the stars appear to move in circles, the angular diameter of these circular orbits being about 41 seconds of arc. This can be understood as follows. Imagine that a star is directly overhead so that a telescope would have to be pointed straight up to see it if the earth were at rest in the ether. That is (see Fig. 1-9a), the rays of light coming from the star would proceed straight down the telescope tube. Now, imagine that the earth is moving to the right through the ether with a speed $v$. In order for the rays to pass down the telescope tube without hitting its sides—that is, in order to see the star—we would have to tilt the telescope as shown in Fig. 1-9b. The light proceeds straight down in the ether (as before) but, during the time $\Delta t$ that the light travels the vertical distance $l = c \, \Delta t$ from the objective lens to the eyepiece, the telescope has moved a distance $v \, \Delta t$ to the right. The eyepiece, at the time the ray leaves the telescope, is on the same vertical line as the objective lens was at the time the ray entered the telescope. From the point of view of the telescope, the ray travels along the axis from objective lens to eyepiece. The angle of tilt of the telescope, $\alpha$, is given by

$$\tan \alpha = \frac{v \, \Delta t}{c \, \Delta t} = \frac{v}{c}. \qquad (1\text{-}11)$$

It was known that the earth goes around the sun at a speed of about 30 km/s so that with $c = 300 \times 10^5$ km/s, we obtain an angle $\alpha = 20.5''$ of arc. This is a very small angle, being about half the angular diameter of the planet Jupiter at its mean distance from the earth. The earth's motion is nearly circular, so the direction of aberration reverses every six months, the telescope axis tracing out a cone of aberration during the year (Fig. 1-9c). The angular diameter of the cone or of the observed circular path of the star, would then be $2\alpha = 41''$ of arc, in excellent agreement with the observations.

The important thing we conclude from this agreement is that the ether is *not* dragged around with the earth. If it were, the ether would be at rest with respect to the earth, the telescope would not have to be tilted, and there would be no aberration at all. That is, the ether would be moving (with the earth) to the right with speed $v$ in Fig. 1-9b, so there would be no need to correct for the earth's motion through the ether, the light ray would be swept along with the ether just as a wind carries a sound wave with it. Hence, *if* there is an ether, it is *not* dragged along by the earth but, instead, the earth moves freely through it. Therefore, we cannot explain the Michelson-Morley result by means of an ether-drag hypothesis.

**The Fizeau Convection Experiment.** One way to investigate whether or not the ether (presuming it exists) is dragged along by bodies moving through it is to

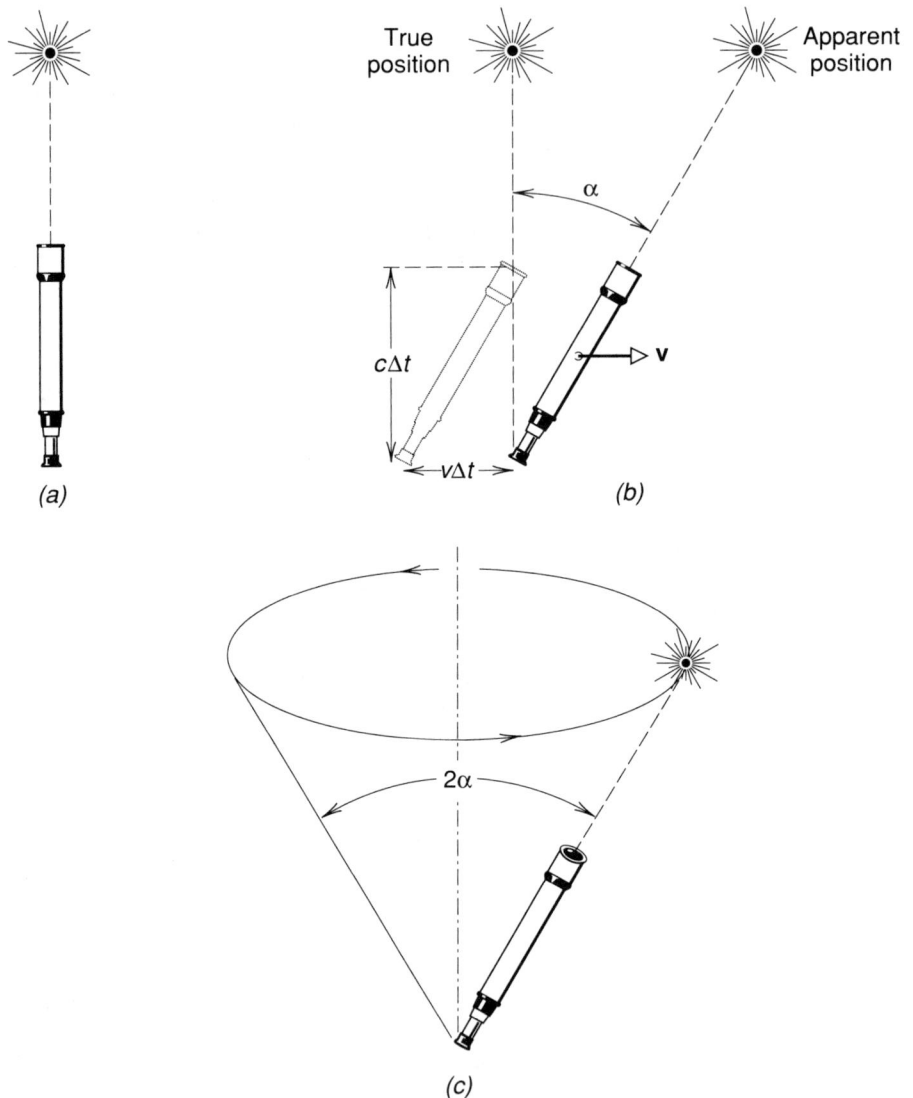

(a)

(b)

(c)

**FIGURE 1-9.** (a) The star and telescope have no relative motion (that is, both are at rest in the ether); the star is directly overhead. (b) The telescope now moves to the right at speed $v$ through the ether; it must be tilted at an angle $\alpha$ (greatly exaggerated in the drawing) from the vertical to see the star, whose apparent position now differs from its true position. ("True" means with respect to the sun, that is, with respect to an earth that has no motion relative to the sun.) (c) A cone of aberration of angular diameter $2\alpha$ is swept out by the telescope axis during the year.

make a direct laboratory test. Assume that a transparent body (water filling a long pipe of length $l$, say) is at rest in the laboratory. The speed at which light is propagated through it is given by $c/n$, where $n$ is the index of refraction of water. Suppose now that the water is caused to move through the pipe at speed $v_w$ with

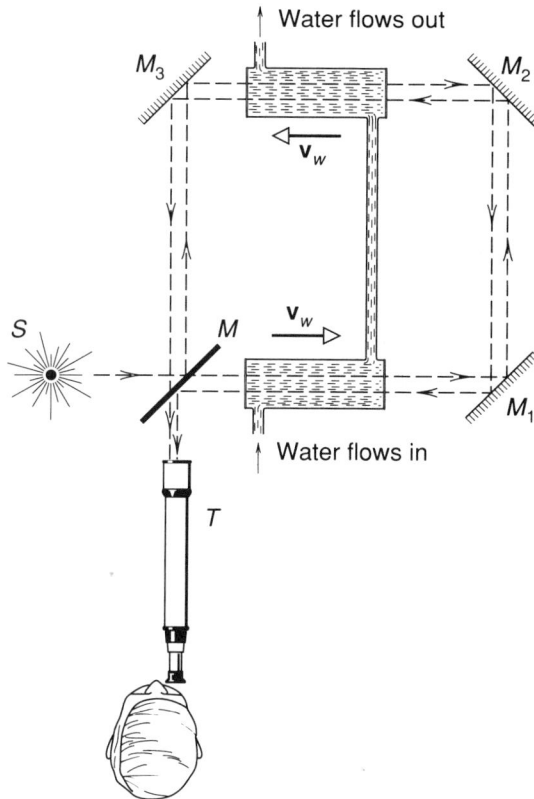

**FIGURE 1-10.** Schematic view of the Fizeau experiment.

respect to the laboratory and that a beam of light is sent along the axis of the pipe, either parallel or antiparallel to the direction of the flow of the water. What will now be the measured laboratory speed of the light going through the (moving) water?

If the moving transparent medium drags the ether with it *totally,* the measured speed of light should be simply $(c/n \pm v_w)$, in which the choice of sign depends on whether the light beam is sent in the direction of motion $(+)$ or opposite to that direction $(-)$.

In 1817 (long before the Michelson-Morley experiment), the French physicist J. A. Fresnel predicted that light would be only *partially* dragged along by a moving medium and derived an exact formula for the effect on the basis of an ether hypothesis. The effect was confirmed experimentally by Fizeau in 1851. The setup of the Fizeau experiment is shown diagrammatically in Fig. 1-10. Light from the source $S$ falls on a partially silvered mirror $M$, which splits the beam into two parts. One part is transmitted to mirror $M_1$ and proceeds in a counterclockwise sense back to $M$, after reflections at $M_1$, $M_2$, and $M_3$. The other part is reflected to $M_3$ and proceeds in a clockwise sense back to $M$, after reflections at $M_3$, $M_2$, $M_1$. At $M$, part of the returning first beam is transmitted and part of the returning second beam is reflected to the telescope $T$. Interference fringes, representing optical path differences of the beams, will be seen in the telescope. Water flows through the tubes (which have flat glass end sections) as shown, so that one light beam always travels in the direction of flow and the other always travels opposite to the direction of flow. The flow of water can be reversed, of course, but outside the tubes conditions remain the same for each beam.

Let the apparatus be our $S$ frame. In this laboratory frame the velocity of light in still water is $c/n$ and the velocity of the water is $v_w$. Does the flow of water, the medium through which the light passes, affect the velocity of light measured in the laboratory? According to Fresnel, the answer is yes. The velocity of light, $v$, in a body of refractive index, $n$, moving with a velocity $v_w$ relative to the observer (that is, to the frame of reference $S$ in which the free-space velocity of light would be $c$) is given by Fresnel as

$$v = \frac{c}{n} \pm v_w \left(1 - \frac{1}{n^2}\right). \qquad (1\text{-}12)$$

The factor $(1 - 1/n^2)$ is called the Fresnel drag coefficient. The speed of light is changed from the value $c/n$ because of the motion of the medium, but because the factor is less than unity, the change (increase or decrease) of speed is less than the speed $v_w$ of the medium—hence the term "drag." For yellow sodium light in water, for example, the speed increase (or decrease) is $0.434 v_w$. Notice that for $n = 1$ ("a moving vacuum"), Eq. 1-12 reduces plausibly to $v = c$.

This result can be understood by regarding the light as being carried along both by the refractive medium and by the ether that permeates it. Then, with the ether at rest and the refractive medium moving through the ether, the light will act to the rest observer as though only a part of the velocity of the medium were added to it. The result can be derived directly from electromagnetic theory.

In Fizeau's experiment, the water flowed through the tubes at a speed of about 7 m/s. Fringe shifts were observed from the zero flow speed to flow speeds of 7 m/s, and on reversing the direction of flow. Fizeau's measurements confirmed the Fresnel prediction. The experiment was repeated by Michelson and Morley in 1886 and by P. Zeeman and others after 1914 under conditions allowing much greater precision, again confirming the Fresnel drag coefficient.

## EXAMPLE 5.

***Fizeau's Ether-Drag Experiment.*** In Fizeau's experiment, the approximate values of the parameters were as follows: $l = 1.5$ m, $n = 1.33$, $\lambda = 5.3 \times 10^{-7}$ m, and $v_w = 7.1$ m/s. A shift of 0.23 fringe was observed from the case $v_w = 0$. Calculate the drag coefficient and compare it with the predicted value.

Let $d$ represent the drag coefficient. The time for beam 1 to traverse the water is then

$$t_1 = \frac{2l}{(c/n) - v_w d}$$

and for beam 2,

$$t_2 = \frac{2l}{(c/n) + v_w d}$$

Hence,

$$\Delta t = t_1 - t_2 = \frac{4l v_w d}{(c/n)^2 - v_w^2 d^2} \cong \frac{4l n^2 v_w d}{c^2}.$$

The period of vibration of the light is $T = \lambda/c$ so that

$$\Delta N = \frac{\Delta t}{T} \cong \frac{4l n^2 v_w d}{\lambda c},$$

and, with the values above, we obtain

$$d = \frac{\lambda c \Delta N}{4l n^2 v_w}$$

$$= \frac{(5.3 \times 10^{-7} \text{ m}) (3.0 \times 10^8 \text{ m/s}) (0.23)}{(4) (1.5 \text{ m}) (1.33)^2 (7.1 \text{ m/s})}$$

$$= 0.49.$$

The Fresnel prediction (see Eq. 1-12) is

$$d = 1 - \frac{1}{n^2} = 1 - \frac{1}{(1.33)^2} = 0.43,$$

the measured value being in rough agreement. Some 35 years later Michelson and Morley re-

peated the experiment with improved techniques and found a value of $0.43 \pm 0.02$ for $d$, in much better agreement with Fresnel's prediction.

If the ether were *totally* dragged along with the water, the speed of light in the laboratory frame would have been, as we have seen, $(c/n + v_w)$ for one beam and $(c/n - v_w)$ for the other beam. Instead, the Fizeau experiment is interpreted most simply in terms of only a *partial* drag by the moving medium. Indeed, the aberration experiment, when done with a telescope filled with water (see Question 21) leads to exactly the same result and interpretation. Hence, the hypothesis that moving bodies are permeated by or closely surrounded by a stagnant localized volume of ether is contradicted by the facts.

At this stage you may well point out that, even though the ether-drag hypothesis is not supported by the facts, Eq. 1-12, derived by Fresnel on the basis of an ether theory, *is* confirmed by experiment. Does not this fact alone amount to support for an ether theory of some kind? In the light of today's understanding, the answer to this question is "no." We shall show in Example 10 in Chapter 2 that Eq. 1-12 follows in a very direct and simple way from the theory of relativity; there is no need whatever to invoke an ether concept in order to derive it.

There appears to be no acceptable experimental basis, then, for the idea of an ether, that is, for a preferred frame of reference. This is true whether we choose to regard the ether as stationary or as dragged along. We must now face the alternative that a principle of relativity is valid in electrodynamics as well as in mechanics. If this is so, then either electrodynamics must be modified, so that it is consistent with the classical relativity principle, or else we need a new relativity principle that is consistent with electrodynamics, in which case classical mechanics will need to be modified.

## 1-8 Attempts to Modify Electrodynamics

Now let us consider attempts to modify the laws of electromagnetism, with the hope of gaining an understanding of the way light is propagated in free space and in transparent bodies. These laws, in the form of Maxwell's equations (see *Physics*, Part II, Sec. 41-8), predict the existence of electromagnetic waves that travel through free space with a speed $c$ ($= 1/\sqrt{\mu_0 \varepsilon_0} = 3.00 \times 10^8$ m/s). A central question is this: With respect to what reference frame does this speed apply? The arguments of the preceding three sections disposed of the ether as such a frame.

In 1908 Walter Ritz discarded the ether concept entirely and modified Maxwell's equations in such a way that the predicted speed $c$ was identified with a reference frame attached to the *source* of the emitted radiation. The various theories that are based on this assumption are called *emission* theories. Common to them all is the hypothesis that the velocity of light is $c$ relative to the original source and that this velocity is independent of the state of motion of the medium transmitting the light.

This would automatically explain the null result of the Michelson-Morley experiment, in which the source is rigidly connected to the interferometer proper and moves with it. The various emission theories differ only in their predictions of the way light is reflected from a moving mirror, a detail that need not involve us here.

It turns out that none of the emission theories can pass the crucial test of experiment. All are contradicted by early experiments using extraterrestrial light sources, such as de Sitter's (1913) observations on double stars and a Michelson-Morley experiment using starlight or sunlight. More recently, measurements of the speed of the radiation emitted from fast-moving energetic particles emerging from laboratory accelerators are even more direct in contradicting emission theories.

Two stars that move in orbits about their common center of mass are called double stars. Imagine the orbits to be circles. Now assume that the velocity of the light by which we see them through the empty space is equal to $c + v_s$, where $v_s$ is the component of the velocity of the source relative to the observer, at the time the light is emitted, along the line from the source to the observer. Then, the time for light to reach the earth from the approaching star would be smaller than that from the receding star. As a consequence, the circular orbits of double stars should appear to be eccentric as seen from earth. But measurements show no such eccentricities in the orbits of double stars observed from earth.

The results are consistent with the assumption that the velocity of the light is independent of the velocity of the source. De Sitter's conclusion was that, if the velocity of light is not equal to $c$ but is equal to $c + kv_s$, then $k$ experimentally must be less than $2 \times 10^{-3}$ (see, however, Ref. 7). In 1977 Brecher [8], observing x-rays emitted by binary stars, claims to have reduced this limit to $2 \times 10^{-9}$.

More direct experiments [9] using fast-moving terrestrial sources confirm the conclusion that the velocity of electromagnetic radiation is independent of the velocity of the source. In such an experiment (1964), measurements were made of the speed of electromagnetic radiation emitted from the decay of rapidly moving $\pi^0$ mesons produced in the CERN synchrotron. The mesons had energies greater than 6 GeV ($v_s = 0.99975c$), and the speed of the $\gamma$ radiation emitted from these fast-moving sources was measured absolutely by timing over a known distance. The result corresponded to a value of $k$ equal to $(-3 \pm 13) \times 10^{-5}$.

The Michelson-Morley experiment, using an extraterrestrial source, has been performed by R. Tomaschek, who used starlight, and D. C. Miller, who used sunlight. If the source velocity (due to rotational and translational motions relative to the interferometer) affects the velocity of light, we should observe complicated fringe-pattern changes. No such effects were observed in either of the experiments.

We saw earlier that an ether hypothesis is untenable. Now we are forced by experiment to conclude further that the laws of electrodynamics are correct and do not need modification. The speed of light (that is, electromagnetic radiation) is the same in all inertial systems, independent of the relative motion of source and observer. Hence, a relativity principle, applicable both to mechanics *and* to electromagnetism, is operating. Clearly, it cannot be the Galilean principle, since that required the speed of light to depend on the relative motion of source and observer. We conclude that the Galilean transformations must be replaced and, therefore, the basic laws of mechanics, which were consistent with those transformations, need to be modified so that they will be consistent with the new transformations.

# 1-9 The Postulates of Special Relativity Theory

In 1905, before many of the experiments we have discussed were actually performed, Albert Einstein (1879–1955) provided a solution to the dilemma facing physics. In his paper "On the Electrodynamics of Moving Bodies" [10], Einstein wrote:

> . . . no properties of observed facts correspond to a concept of absolute rest; . . . for all coordinate systems for which the mechanical equations hold, the equivalent elec-trodynamical and optical equations hold also. . . . In the following we make these assumptions (which we shall subsequently call the Principle of Relativity) and intro-duce the further assumption—an assumption which is at the first sight quite irrecon-cilable with the former one—that light is propagated in vacant space, with a velocity $c$ which is independent of the nature of motion of the emitting body. These two assumptions are quite sufficient to give us a simple and consistent theory of electro-dynamics of moving bodies on the basis of the Maxwellian theory for bodies at rest.

We can rephrase these assumptions of Einstein as follows.

1. *The laws of physics are the same in all inertial systems. No preferred inertial system exists* (the principle of relativity).
2. *The speed of light in free space has the same value c in all inertial systems* (the principle of the constancy of the speed of light).

Einstein's relativity principle goes beyond the Newtonian relativity principle, which dealt only with the laws of mechanics, to include *all* the laws of physics. It states that it is impossible by means of *any* physical measurements to designate an inertial system as intrinsically stationary or moving; we can speak only of the *relative* motion of two systems. Hence, no physical experiment of *any* kind made entirely *within* an inertial system can tell the observer what the motion of his system is with respect to any other inertial system.

Although the relativity principle seems quite acceptable to us (and, in retrospect, even compelling), there are conceptual difficulties for most of us in accepting Einstein's second postulate. Consider that a source emits a short pulse of light whose spherical wavefronts sweep past three observers, *A, B,* and *C*. Let *A* be at rest with respect to the light source; let *B* be rushing toward the source at great speed; finally, let *C* be rushing away from the source, also at great speed. The principle of the constancy of the speed of light tells us that *all three observers will measure the same speed for the light pulse!* This is in flat contradiction to the predictions of the Galilean velocity transformation equations (Eq. 1-5). It is *not* the way sound pulses or baseballs behave. It is a tribute to Einstein that he first realized clearly that— conceptual difficulties aside—this *is* the way that light behaves.

The entire special theory of relativity is derived directly from the two assump-tions stated above. Their simplicity, boldness, and generality are characteristic of Einstein's genius. The success of his theory, as indeed of any theory, can be judged only by comparison with experiment. It was able not only to explain all the existing experimental results but also to predict new effects that were confirmed by later experiments. No experimental objection to Einstein's special theory of relativity has yet been found.

In Table 1-2 we list the seven theories proposed at various times and compare their predictions of the results of 13 crucial experiments, old and new. Notice that only the special theory of relativity is in agreement with *all* the experiments listed. We have already commented on the successes and failures of the ether and emission theories with most of the light-propagation experiments, and it remains for us to show how special relativity accounts for their results. In addition, several experiments from other fields—some suggested by the predictions of relativity and in flat contradiction to Newtonian mechanics—remain to be examined. What emerges from this comparative preview is the compelling *experimental* basis of special relativity theory. It alone is in accord with the real world of experimental physics.

As is often true in the aftermath of a great new theory, it seemed obvious to many in retrospect that the old ideas had to be wrong. For example, Herman Bondi has written:

> The special theory of relativity is a necessary consequence of any assertion that the unity of physics is essential, for it would be intolerable for all inertial systems to be equivalent from a dynamical point of view yet distinguishable by optical measurements. It now seems almost incredible that the possibility of such a discrimination was taken for granted in the nineteenth century, but at the same time it was not easy to see what was more important—the universal validity of the Newtonian principle of relativity or the absolute nature of time.

It was his preoccupation with the nature of time that led Einstein to his revolutionary proposals. We shall see later how important a clear picture of the concept of time was to the development of relativity theory. However, the program of the theory, in terms of our discussions in this chapter, should now be clear. First, we must obtain equations of transformation between two uniformly moving (inertial) systems that will keep the velocity of light constant. Second, we must examine the laws of physics to check whether or not they keep the same form (that is, are invariant) under this transformation. Those laws that are not invariant will need to be generalized so as to obey the principle of relativity.

The new equations of transformation obtained in this way by Einstein are known for historical reasons as a Lorentz transformation. We have seen (Section 1-3) that Newton's equation of motion is invariant under a Galilean transformation, which we now know to be incorrect. It is likely then that Newton's laws—and perhaps other commonly accepted laws of physics—will not be invariant under a Lorentz transformation. In that case, they must be generalized. We expect the generalization to be such that the new laws will reduce to the old ones for velocities much less than that of light, for in that range both the Galilean transformation and Newton's laws are at least approximately correct.

## 1-10 Einstein and the Origin of Relativity Theory

It is so fascinating a subject that one is hard-pressed to cut short a discussion of Albert Einstein, the person. Common misconceptions of the man, who quite prop-

**TABLE 1-2**
Experimental Basis for the Theory of Special Relativity[a]

| Theory | | Light Propagation Experiments | | | | | | | Experiments from Other Fields | | | | | |
|---|---|---|---|---|---|---|---|---|---|---|---|---|---|---|
| | | Aberration | Fizeau Convection Coefficient | Michelson-Morley | Kennedy-Thorndike | Moving Sources and Mirrors | De Sitter Spectroscopic Binaries | Michelson-Morley, Using Sunlight | Variation of Mass with Velocity | General Mass–Energy Equivalence | Radiation from Moving Charges | Meson Decay at High Velocity | Trouton-Noble | Unipolar Induction, Using Permanent Magnet |
| Ether Theories | Stationary Ether, No Contraction | √ | √ | × | × | √ | √ | × | × | ▦ | √ | ▦ | × | × |
| | Stationary Ether, Lorentz Contraction | √ | √ | √ | × | √ | √ | √ | √ | ▦ | √ | ▦ | √ | × |
| | Ether Attached to Ponderable Bodies | × | × | √ | √ | √ | √ | √ | × | ▦ | ▦ | ▦ | √ | ▦ |
| Emission Theories | Original Source | √ | √ | √ | √ | √ | × | × | ▦ | ▦ | × | ▦ | ▦ | ▦ |
| | Ballistic | √ | ▦ | √ | √ | × | × | × | ▦ | ▦ | × | ▦ | ▦ | ▦ |
| | New Source | √ | ▦ | √ | √ | × | × | √ | ▦ | ▦ | × | ▦ | ▦ | ▦ |
| Special Theory of Relativity | | √ | √ | √ | √ | √ | √ | √ | √ | √ | √ | √ | √ | √ |

*Legend.* √, the theory agrees with experimental results.
    ×, the theory disagrees with experimental results.
    ▦, the theory is not applicable to the experiment.
[a]From Ref. 2.

erly symbolized for his generation the very height of intellect, might be shattered by such truths as these: Einstein's parents feared for a while that he might be mentally retarded, for he learned to speak much later than customary; one of his teachers said to him, "You will never amount to anything, Einstein," in despair at his daydreaming and his negative attitude toward formal instruction; he failed to get a high-school diploma and, with no job prospects, at the age of 15 he loafed like a "model drop-out"; Einstein's first attempt to gain admission to a polytechnic institute ended when he failed to pass an entrance examination; after gaining admittance he cut most of the lectures and, borrowing a friend's class notes, he crammed intensively for two months before the final examinations. He later said of this ". . . after I had passed the final examination, I found the consideration of any scientific problem distasteful to me for an entire year." It was not until two years after his graduation that he got a steady job, as a patent examiner in the Swiss Patent Office at Berne; Einstein was very interested in technical apparatus and instruments, but—finding that he could complete a day's work in three or four hours—he secretly worked there, as well as in his free time, on the problems in physics that puzzled him. And so it goes.*

The facts above are surprising only when considered in isolation, of course, Einstein simply could not accept the conformity required of him, whether in educational, religious, military, or governmental institutions. He was an avid reader who pursued his own intellectual interests, had a great curiosity about nature, and was a genuine "free thinker" and independent spirit. As Martin Klein points out, what is really surprising about Einstein's early life is that none of his "elders" recognized his genius.

But such matters aside, let us look now at Einstein's early work. It is appropriate to quote here from Martin Klein [13].

> In his spare time during those seven years at Berne, the young patent examiner wrought a series of scientific miracles; no weaker word is adequate. He did nothing less than to lay out the main lines along which twentieth-century theoretical physics has developed. A very brief list will have to suffice. He began by working out the subject of statistical mechanics quite independently and without knowing of the work of J. Willard Gibbs. He also took this subject seriously in a way that neither Gibbs nor Boltzman had ever done, since he used it to give the theoretical basis for a final proof of the atomic nature of matter. His reflections on the problems of the Maxwell-Lorentz electrodynamics led him to create the special theory of relativity. Before he left Berne he had formulated the principle of equivalence and was struggling with the problems of gravitation which he later solved with the general theory of relativity. And, as if these were not enough, Einstein introduced another new idea into physics, one that even he described as "very revolutionary," the idea that light consists of particles of energy. Following a line of reasoning related to but quite distinct from Planck's, Einstein not only introduced the light quantum hypothesis, but proceeded almost at once to explore its implications for phenomena as diverse as photochemistry and the temperature dependence of the specific heat of solids.
>
> What is more, Einstein did all this completely on his own, with no academic connections whatsoever, and with essentially no contact with the elders of his profession.

* See Refs. 11–19 for some rewarding articles and books about Einstein.

The discussion thus far emphasizes Einstein's independence of other contemporary workers in physics. Also characteristic of his work is the fact that he always made specific predictions of possible experiments to verify his theories. In 1905, at intervals of less than eight weeks, Einstein sent to the *Annalen der Physik* three history-making papers. The first paper, on the quantum theory of light, included an explanation of the photoelectric effect. The suggested experiments, which gave the proof of the validity of Einstein's equations, were successfully carried out by Robert A. Millikan nine years later! The second paper, on statistical aspects of molecular theory, included a theoretical analysis of Brownian movement. Einstein wrote later of this: "My major aim in this was to find facts which would guarantee as much as possible the existence of atoms of definite size. In the midst of this I discovered that, according to atomistic theory, there would have to be a movement of suspended microscopic particles open to observation, without knowing that observations concerning the Brownian motion were already long familiar." Einstein's specific predictions were confirmed in detail in 1908 by Jean Perrin. The third paper [10], on special relativity, included applications to electrodynamics such as the relativistic mass of a moving body, all subsequently confirmed by experiment.

As for the origins of his special theory of relativity, Einstein was probably convinced, independently of the experimental background that we have presented, that Maxwell's equations had to have exactly the same form in all inertial frames of reference. He may have reached such a conclusion intuitively in his youth while meditating on how a light wave would look to an observer who was following it at speed $c$. The constancy of the velocity of light seemed to him indisputable, and his 1905 paper begins with the precise electrodynamical experimental situation described in a text on Maxwell's theory that he was studying at the time. This highlights another characteristic of Einstein's work, which suggests why his approach to a problem was usually not that of the mainstream: namely, his attempt to restrict hypotheses to the smallest number possible and to the most general kind. For example, Lorentz, who never really accepted Einstein's relativity, used a great many *ad hoc* hypotheses to arrive at the same transformations in 1904 as Einstein did in 1905 (and as Voight did in 1887); furthermore, Lorentz had assumed these equations *a priori* in order to obtain the invariance of Maxwell's equations in free space. Einstein, on the other hand, *derived* them from the simplest and most general postulates—the two fundamental principles of special relativity. And he was guided by his solution to the problem that had occupied his thinking since he was 16 years old: the nature of time. Lorentz and Poincaré, two of the most eminent mathematical physicists of the time, had accepted Newton's universal time ($t = t'$), whereas Einstein abandoned that notion.

Newton, even more than many succeeding generations of scientists, was aware of the fundamental difficulties inherent in his formulation of mechanics, based as it was on the concepts of absolute space and absolute time. Einstein expressed a deep admiration for Newton's method and approach and can be regarded as bringing many of the same basic attitudes to bear on his analysis of the problem. In his Autobiographical Notes, after critically examining Newtonian mechanics, Einstein writes:

Enough of this. Newton, forgive me; you found the only way which, in your age, was just about possible for a man of highest thought and creative power. The concepts, which you created, are even today still guiding our thinking in physics, although we now know that they will have to be replaced by others farther removed from the sphere of immediate experience, if we aim at a profounder understanding of relationships.

It seems altogether fitting that Einstein should have extended the range of Newton's relativity principle, generalized Newton's laws of motion, and later incorporated Newton's law of gravitation into his space-time scheme. In subsequent chapters we shall see how this was accomplished.

## QUESTIONS

1. Quasars (*quasi*-stell*ar* objects) are the most intrinsically luminous objects in the universe. Many of them fluctuate in brightness, often on a time scale of a day or so. How can the rapidity of these brightness changes be used to estimate an upper limit to the size of these objects? (*Hint:* Separated points cannot change in a coordinated way unless information is sent from one to the other.)

2. How would you test a proposed reference frame to find out whether or not it is an inertial frame?

3. Give several examples of reference frames that are not inertial frames.

4. If you were confined to the hold of an ocean liner on a calm sea, how could you test whether the ship was moving with a constant velocity? Was accelerating?

5. Give examples in which effects associated with the earth's rotation are significant enough in practice to rule out a laboratory frame as being a good enough approximation to an inertial frame.

6. How does the concept of simultaneity enter into the measurement of the length of a body?

7. Could a mechanical experiment be performed in a given reference frame that would reveal information about the *acceleration* of that frame relative to an inertial frame?

8. In an inelastic collision, the amount of thermal energy (internal mechanical kinetic energy) developed is independent of the inertial reference frame of the observer. Explain why, in words.

9. Why is it necessary to rotate the interferometer in the Michelson-Morley experiment?

10. You are in deep space, in command of a spaceship equipped with all the resources of a modern optics laboratory, including a Michelson interferometer. Taking full advantage of the maneuverability of your ship, how would you go about testing for the presence or absence of a presumed luminiferous ether?

11. Describe an acoustic Michelson-Morley experiment, in analogy with the optical one. What could you measure with such a device? Would you expect a null result, similar to that obtained in the optical case?

12. A simple way to test for the presence or absence of an ''ether wind'' would be to make one-way, rather than round-trip, measurements of the speed of light.

That is, we could measure the speed along a straight line, first in one direction and then in the other. Explain how the speed of the ether wind could be deduced from such data. Is the experiment practical?

13. Does the Lorentz-Fitzgerald contraction hypothesis contradict the classical notion of a rigid body?

14. Could the Lorentz-Fitzgerald contraction hypothesis have been disproved by simply measuring the length of a rod at different orientations to the presumed ether wind?

15. What is the relation between the apparent angular position of the star in Fig. 1-9 and the apparent direction of motion of raindrops falling on the vertical windshield of a bus?

16. Does the fact that stellar aberration is observable contradict the principle of the relativity of uniform motion (that is, does it determine an absolute velocity)? How, in this regard, does it differ from the Michelson-Morley experiment?

17. In the text we assumed that the star whose stellar aberration we were measuring was directly overhead. How would the situation change if the star were close to the horizon?

18. Even if the speed of light were infinite, the star in Fig. 1-9 would appear to move in a small circle during the course of a year, due to the different directions from which it is viewed from different parts of the earth's orbit. This effect is called *parallax*. How can effects due to parallax be separated from those due to aberration?

19. If the earth's motion, instead of being nearly circular about the sun, were uniformly along a straight line through the "ether," could an aberration experiment measure its speed?

20. How can we use aberration observations to refute the Ptolemaic, or earth-centered, model of the solar system?

21. If the "ether" were dragged along with water, what would be the expected result of the aberration experiment when done with a telescope filled with water? (Such an experiment was done by Sir George Airy in 1871. The results were the same as without water.)

22. Describe in your own words the essential difference between the Michelson-Morley and the Kennedy-Thorndike experiments.

23. Borrowing two phrases from Herman Bondi, we can catch the spirit of Einstein's two postulates by labeling them: (1) the principle of "the irrelevance of velocity" and (2) the principle of "the uniqueness of light." In what senses are velocity irrelevant and light unique in these two statements?

24. A beam from a laser falls at right angles on a plane mirror and rebounds from it. What is the speed of the reflected beam if the mirror is (a) fixed in the laboratory? (b) Moving directly toward the laser with speed $v$?

25. Comment on the assertion that if one accepts Einstein's principle of the constancy of the speed of light, then there is no reason to interpret the null result of the Michelson-Morley experiment as evidence against an ether hypothesis.

26. What boxes in Table 1-2 have been accounted for in this chapter?

27. Show that the two postulates of the special theory of relativity account satisfactorily for the results of (a) the Michelson-Morley experiment, (b) the Kennedy-Thorndike experiment, (c) the stellar aberration observations, (d) the binary star observations, and (e) the CERN experiment described in Section 1-8.

28. The speed of light in a vacuum is a true constant of nature, independent of the wavelength of the light or the choice of an (inertial) reference frame. Is there any sense, then, in which Einstein's second postulate can be viewed as contained within the scope of his first postulate?

29. Discuss the problem that young Einstein grappled with; that is, what would be the appearance of an electromagnetic wave to a person running along with it at speed *c?*

30. Can a particle move through a (transparent) medium at a speed greater than the speed of light in that medium? (See *Physics,* Part II, Sec. 43-4.)

## PROBLEMS*

1. **Some lengths and some times.** (a) How many meters are there in a light-day? (b) How many kilometers in a light-microsecond? (c) How many years in a light-parsec? (d) How many seconds in a light-fermi? A *parsec* ($= 3.09 \times 10^{16}$ m) is a length unit much used in astronomy. A *fermi* ($= 10^{-15}$ m) is a length unit much used in nuclear and particle physics.

2. **Some speeds.** What fraction of the speed of light does each of the following speeds represent? (a) A typical rate of continental drift (1 inch per year, say). (b) A typical drift speed for electrons in a current-carrying conductor (0.5 mm/s, say). (c) A typical highway speed limit of 55 mi/h. (d) The root-mean-square speed of a hydrogen molecule at room temperature. (e) A supersonic plane flying at Mach 2.5 ($= 1200$ km/h, say). (f) The escape speed of a projectile from the surface of the earth. (g) The speed of the earth in its orbit around the sun. (h) The recession speed of a distant quasar ($3.0 \times 10^4$ km/s, say).

3. **Some units for *c*—useful and otherwise.** Express the speed of light in (a) cm/ns, (b) ft/μs, (c) ly/y, (d) $J^{1/2} \cdot kg^{-1/2}$, (e)

MeV$^{1/2} \cdot$ u$^{-1/2}$, and (f) m $\cdot$ F$^{-1/2} \cdot$ H$^{-1/2}$. (Hint: Consider the relations $E = mc^2$ and $c = 1/\sqrt{\mu_0\varepsilon_0}$. In the above "ly" stands for light year and "u" for atomic mass unit.)

4. **Win, place, and show.** Electrons emerging from the 2-mile-long Stanford Electron Accelerator may have a kinetic energy of 20 GeV, corresponding to an electron speed that is less than the speed of light by only 9.78 cm/s. Consider a three-way race over a straight 10-km course between the electron beam in a vacuum, a light beam in a vacuum, and a light beam in air. (a) List the order of finish. (b) Give the time interval between first and second place and (c) between second and third place. Take the index of refraction of air to be 1.00029.

5. **". . . this goodly frame, the earth . . ." (Hamlet).** For some purposes the earth cannot be taken as a totally "goodly" inertial frame because its motion is accelerated. Calculate the accelerations associated with (a) the earth's rotation about its axis (assume an equatorial point), (b) the earth's orbital motion about the sun and (c) the orbital motion of the solar system about the galactic center. The sun is 10 kpc from the galactic

* The more difficult problems have been starred ★.

center and orbits around it at 300 km/s. One parsec = 1 pc = $3.09 \times 10^{16}$ m.

6. **Electrons also fall.** Quite apart from effects due to the earth's rotational and orbital motions, a laboratory frame is not strictly an inertial frame because a particle placed at rest there will not, in general, remain at rest; it will fall under gravity. Often, however, events happen so quickly that we can ignore free fall and treat the frame as inertial. Consider, for example, a 1.0-MeV electron (for which $v = 0.992c$) projected horizontally into a laboratory test chamber and moving through a distance of 20 cm. (a) How long would it take, and (b) how far would the electron fall during this interval? What can you conclude about the suitability of the laboratory as an inertial frame in this case?

7. **The Galilean transformation generalized.** Write the Galilean coordinate transformation equations (see Eqs. 1-1a and 1-1b for the case of an arbitrary direction for the relative velocity **v** of one frame with respect to the other. Consider that the corresponding axes of the two frames remain parallel. (*Hint:* Let **v** have components $v_x, v_y,$ and $v_z$.)

8. **Momentum is conserved for all. . . .** An observer on the ground watches a collision between two particles whose masses are $m_1$ and $m_2$ and finds, by measurement, that momentum is conserved. Use the classical velocity addition theorem (Eq. 1-5) to show that an observer on a moving train will also find that momentum is conserved in this collision.

9. **(. . .but only if mass is conserved!)** Repeat Problem 8 under the assumption that a transfer of mass from one particle to the other takes place during the collision, the initial masses being $m_1$ and $m_2$ and the final masses being $m_1'$ and $m_2'$. Again, assume that the ground observer finds, by measurement, that momentum is conserved. Show that the train observer will also find that momentum is conserved *only* if mass is also conserved, that is, if

$$m_1 + m_2 = m'_1 + m'_2.$$

★10. **The invariance of "elastic."** A collision between two particles in which kinetic energy is conserved is described as *elastic*. Show, using the Galilean velocity transformation equations, that if a collision is found to be elastic in one inertial reference frame, it will also be found to be elastic in all other such frames. Could this result have been predicted from the conservation of energy principle?

11. **The work–energy theorem holds in all inertial frames.** Observer $G$ is on the ground and observer $T$ is on a train moving with uniform velocity **v** with respect to the ground. Each observes that a particle of mass $m$, initially at rest with respect to the train, is acted on by a constant force **F** applied to it in the forward direction for a time $t$. (a) Show that the two observers will find, for the work done on the particle by force **F**,

$$W_T = \tfrac{1}{2}ma^2t^2 \quad \text{and} \quad W_G = \tfrac{1}{2}ma^2t^2 + mvat,$$

in which $a$ is the common acceleration of the particle. (b) Show that $\Delta K_T$ and $\Delta K_G$, the changes in kinetic energy calculated by each observer, are also given by these same two expressions. Thus the work–energy theorem ($W = \Delta K$) is valid in all inertial reference frames.

12. **More work–energy.** (a) In Problem 11, evaluate $W_T$ and $W_G$ for $m = 1.0$ kg, $a = 0.20g$, $t = 5.0$ s, and $v = 25$ m/s. (b) Explain the fact that the two observers find that different amounts of work are done by the same force in terms of the different distances through which the observers measure that force to act during the time $t$. (c) Find also the initial and final kinetic energies for the particle, as reported by each observer. (d) Verify that the work–energy theorem is valid in each frame. (e) Explain the fact that the observers find different final kinetic energies for the particle in terms of the work that the particle could do in being brought to rest relative to each observer's frame.

13. **Work, friction, and thermal energy.** Suppose that, in Problem 11, the "particle of mass $m$" is a block sliding on the floor of the train and that a frictional force acts between the block and the floor. Assume that the external applied force acts for the same time $t$ and is of such a magnitude that the block experiences the same acceleration $a$ as in that problem. (a) Show that the amount of thermal energy developed by the action of the frictional force is the same for each observer. (*Hint:* Work done against friction depends on the *relative* motion of the surfaces.) (b) Calculate this thermal energy, assuming that the frictional force is equal to $0.10mg$ and that, as in Problem 12, $a = 0.20g$, $m = 1.0$ kg, and $t = 5.0$ s.

14. **Steve and Sally watch an elastic collision.** (a) Observer Steve watches a head-on elastic collision, in which one of the colliding particles is initially at rest in his reference frame. The masses and the initial velocities are shown in the table below. Fill in the blanks in Steve's column in this table. Verify numerically that momentum and kinetic energy are indeed conserved. (b) Observer Sally, moving past Steve at 2.5 m/s along the collision line in the same direction as $m_1$ also sees this event. Using the Galilean transformation equations, fill in the blanks in *her* column and, again, verify that momentum and kinetic energy are conserved.

15. **Steve and Sally watch an inelastic collision.** The two students of Problem 14 now observe a second head-on collision between two particles. In this case, however, it is *not* assumed that kinetic energy is conserved; instead, the final velocity of the particle of mass $m_2$ is given. Fill in the blanks, (a) for Steve and (b) for Sally, as before. (c) Do

**TABLE 1-3**
Problem 14

| | Symbol[a] | Steve's Data | Sally's Data |
|---|---|---|---|
| Particle Masses, kg | $m_1$ | 0.107 | 0.107 |
| | $m_2$ | 0.345 | 0.345 |
| Particle Velocities, m/s | $u_1$ | +3.25 | |
| | $u_2$ | 0.00 | |
| | $u'_1$ | | |
| | $u'_2$ | | |
| Total System Momentum, kg · m/s | $P$ | | |
| | $P'$ | | |
| | $P' - P$ | | |
| Total System Kinetic Energy, J | $K$ | | |
| | $K'$ | | |
| | $K' - K$ | 0.000 | |

[a]Primed quantities refer to values after the collision.

**TABLE 1-4**
Problem 15

| | Symbol[a] | Steve's Data | Sally's Data |
|---|---|---|---|
| Particle Masses, kg | $m_1$ | 0.107 | 0.107 |
| | $m_2$ | 0.345 | 0.345 |
| Particle Velocities, m/s | $u_1$ | +3.25 | |
| | $u_2$ | 0.00 | |
| | $u'_1$ | | |
| | $u'_2$ | +1.25 | |
| Total System Momentum, kg · m/s | $P$ | | |
| | $P'$ | | |
| | $P' - P$ | | |
| Total System Kinetic Energy, J | $K$ | | |
| | $K'$ | | |
| | $K' - K$ | | |

[a]Primed quantities refer to values after the collision.

Steve and Sally still agree that momentum is conserved? Do they agree that kinetic energy is *not* conserved? Do they find the same value for the decrease in kinetic energy? Where has this energy gone?

16. **Maxwell versus Galileo.** Starting from Maxwell's equations, it is possible to derive a *wave equation* whose solutions represent electromagnetic waves. For the special case of a plane wave traveling parallel to the $x$ axis in the direction of increasing $x$, the wave equation proves to be

$$\frac{\partial^2 E}{\partial x^2} = \frac{1}{c^2} \frac{\partial^2 E}{\partial t^2}$$

in which $E(x, t)$ is the wave amplitude. Show that this equation does not retain its form when written in terms of the coordinates $(x', t')$ appropriate to another reference frame; That is, show that this equation is *not* invariant under a Galilean transformation. *Hint:* Use the chain rule in which, if $f = f(x, t)$, then

$$\frac{\partial f}{\partial x} = \frac{\partial f}{\partial x'}\left(\frac{\partial x'}{\partial x}\right) + \frac{\partial f}{\partial t'}\left(\frac{\partial t'}{\partial x}\right).$$

17. **The "binumeral explosion" (one student's preference).** (a) Use the binomial expansion theorem to verify each of the expansions displayed in the footnote on page 16. (b) Assuming that $\beta = 0.30$, what percent error is made in the first of these expansions if only two terms are kept? Only three terms? (c) Answer these same questions for the second of these expansions.

★18. **Michelson-Morley, generalized.** Figure 1-11 shows a Michelson interferometer (compare Fig. 1-4) with a presumed ether wind whose uniform velocity **v** makes an angle $\phi$ with the arm of length $l_1$. (a) Show that the round-trip travel times for the two arms are

$$t_1 = \frac{2l_1}{c} \frac{\sqrt{1 - \beta^2 \sin^2 \phi}}{1 - \beta^2} \quad \text{and}$$

$$t_2 = \frac{2l_2}{c} \frac{\sqrt{1 - \beta^2 \cos^2 \phi}}{1 - \beta^2}.$$

(b) Let the interferometer be rotated through 90°, starting from the position shown in the

figure. Show that the predicted fringe shift is given by

$$\Delta N = \frac{\beta^2(l_1 + l_2)}{\lambda} \cos 2\phi.$$

(c) Show that the three expressions given in (a) and (b) reduce to those given in Section 1-5 for $\phi = 0$ or 90° and, further, that the predictions for $\phi = 45°$ are reasonable.

19. **Michelson-Morley with a real wind.** A pilot plans to fly due east from $A$ to $B$ and back again. If $u$ is his airspeed and if $l$ is the distance between $A$ and $B$, it is clear that his round-trip time $t_0$—if there is no wind—will be $2l/u$. (a) Suppose, however, that a steady wind of speed $v$ blows from the east (or from the west). Show that the round-trip travel time will now be

$$t_1 = \frac{t_0}{1 - (v/u)^2}.$$

(b) If the wind is from the north (or from the south), show that the expected round-trip travel time is

$$t_2 = \frac{t_0}{\sqrt{1 - (v/u)^2}}.$$

(c) Note that these two travel times are *not* equal. In the Michelson-Morley experiment, however, experiment seems to show that (for arms of equal length) the travel times for light *are* equal; otherwise these experiment-

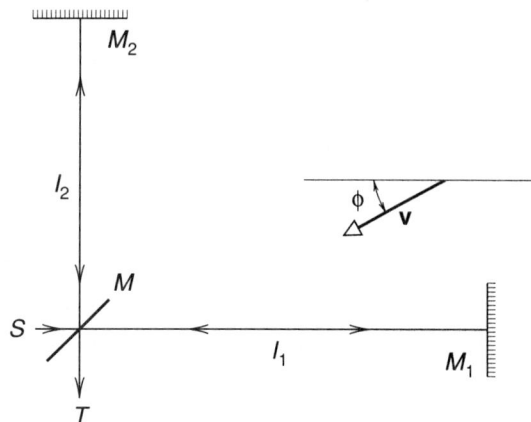

**FIGURE 1-11.** Problem 18.

ers would have found a fringe shift when they rotated the interferometer. What is the essential difference between the two situations?

20. **A Michelson-Morley derby.** Two light planes, whose cruising airspeeds are 100 knots (= 115 mi/h), each complete a round trip over a 40-km course laid out on the ground. The two courses are at right angles, and if there is no wind the race will be a tie. Suppose, however, that there is a steady 15-knot wind parallel to one of the tracks. (a) What heading with respect to her ground track must the cross-wind pilot adopt? (b) Which pilot will win and by how much?

21. **Not much of a contraction.** In the Michelson-Morley experiment, by about how many atomic diameters would the appropriate arm of the interferometer have to shrink according to the Lorentz-Fitzgerald contraction hypothesis? The actual arm length of the interferometer was 2.8 m, but (see Fig. 1-8) the beam was reflected back and forth four times on each arm. Take 0.1 nm as a typical atomic diameter.

22. **Still not much!** If the Lorentz-Fitzgerald contraction hypothesis were true, by how much would the diameter of the earth be contracted by virtue of its orbital motion?

23. **A neat little result.** Suppose that a rod in the form of a circular cylinder is immersed in a presumed ether wind, the wind velocity being at right angles to the axis of the rod. Show that the effect of the Lorentz-Fitzgerald contraction would be to change the cross section of the rod from a circle to an ellipse. Show further that the eccentricity of this ellipse is simply $\beta$.

★24. **The incredible shrinking rod.** Figure 1-12 shows a rod placed at an angle $\phi$ to the direction of an ether wind of velocity **v**. (a) Show that, according to the Lorentz-Fitzgerald contraction hypothesis, the length of the rod is given by

$$l = l^0 \sqrt{1 - \beta^2 \cos^2 \phi},$$

in which $\beta = v/c$ and $l^0$ is the length of the rod when there is no ether wind. (b) Show further that according to the Lorentz-Fitzgerald hypothesis the round-trip travel time for a light signal sent back and forth along the rod is given, to order $\beta^2$, by

$$t = \frac{2l^0}{c}\left(1 + \frac{1}{2}\beta^2\right).$$

(*Hint:* Make use of the expression for $t_1$ displayed in Problem 18.) Note that the round-trip travel time is independent of $\phi$. What is the relationship of this problem to the Michelson-Morley experiment? Discuss this problem from the point of view of Einstein's second postulate.

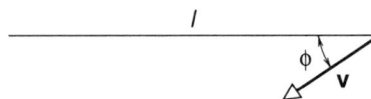

**FIGURE 1-12**   Problem 24.

★25. **A fascinating "thought experiment."** Paul Ehrenfest (1880–1933) proposed the following thought experiment to illustrate the different behavior expected for light under the ether-wind hypothesis and under Einstein's second postulate.

Imagine yourself seated at the center of a spherical shell of radius $3 \times 10^8$ m, the inner surface being diffusely reflecting. A source at the center of the sphere emits a sharp pulse of light, which travels outward through the darkness with uniform intensity in all directions. What would you see during the 3-s interval following the pulse under the assumptions that (a) there is a steady ether wind blowing through the sphere at 100 km/s and (b) that there is no ether and Einstein's second postulate holds. (c) Discuss the relationship of this thought experiment to the Michelson-Morley experiment.

26. **Aberration and the rotating earth.** In addition to revolving about the sun, the earth also rotates on its axis once a day. Find the largest aberration angle (tilt of the telescope)

due to the earth's rotation alone for an observer at (a) the equator, (b) latitude 60°, and (c) the north pole.

27. **Aberration and relativity.** Consider the aberration arrangement of Fig. 1-9b. (a) At what speed does light pass along the telescope axis according to the ether hypothesis? (b) According to the special theory of relativity? (c) Show that, according to relativity theory, the classical aberration equation (see Eq. 1-11)

$$\tan \alpha = \frac{v}{c} \quad \text{(classical theory)}$$

must be replaced by

$$\sin \alpha = \frac{v}{c} \quad \text{(relativity theory)}$$

Thus the ether theory and the theory of relativity make different predictions for the aberration of starlight. As the next problem shows, however, the predictions prove to be not as different as they seem.

28. **Not a very sensitive test.** Problem 27 displays the classical and the relativistic predictions for the aberration constant $\alpha$. If $\alpha_c$ and $\alpha_r$ are the predictions of these two theories, find their fractional difference; that is, find $(\alpha r - ac)/\alpha r$. *Assume that v*, the earth's orbital speed, is 30 km/s, and take $3.00 \times 10^8$ m/s for the speed of light. (*Hint:* The predictions of the two theories are so close that the calculation of their difference may well be beyond the scope of your hand calculator! Use the series expansions:

$$\sin^{-1} x = x + \tfrac{1}{6}x^3 + \tfrac{3}{40}x^5 \cdots$$

and

$$\tan^{-1} x = x - \tfrac{1}{3}x^3 + \tfrac{1}{5}x^5 \cdots).$$

29. **Weighing the sun.** (a) Show that $M$, the mass of the sun, is related to the aberration constant $\alpha$ (see Eq. 1-11) by

$$M = \frac{\alpha^2 c^2 R}{G},$$

in which $R$ is the radius of the earth's orbit (assumed circular) and $G(= 6.67 \times 10^{-11}$ N·m²/kg²) is the universal gravita-

tional constant. (*Hint:* Apply Newton's second law to the earth's motion around the sun.) (b) Calculate $M$, given that $\alpha = 20.5''$ and $R = 1.50 \times 10^{11}$ m.

30. **Dragging the ether.** (a) In the Fizeau ether-drag experiment (Fig. 1-10), identify the frames $S$ and $S'$ and the relative velocity $v$ that correspond to Fig. 1-1. (b) Show that in the Fresnel ether-drag formula (Eq. 1-12) $v \rightarrow v_w$ for very large values of $n$. How would you interpret this? (c) Under what circumstances would the Fresnel drag coefficient be zero? To what does this correspond physically?

31. **A fast-moving fluid.** According to the Fresnel ether-drag formula (Eq. 1-12), with what speed would the fluid have to move in order that the speed of light in the fluid equal the speed of light in a vacuum? Assume an index of refraction of $n = 1.33$.

32. **Twinkle, twinkle, double star. . . .** Consider one star in a binary system moving in uniform circular motion with speed $v$. Consider two positions: (I) the star is moving *away* from the earth along the line connecting them, and (II) the star is moving *toward* the earth along the line connecting them (see Fig. 1-13). Let the period of the star's motion be $T$ and its distance from the earth be $l$. Assume that $l$ is large enough that positions

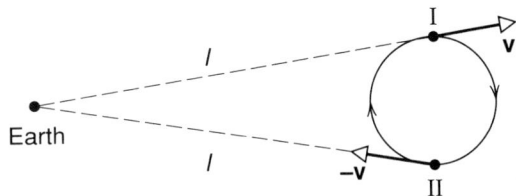

**FIGURE 1-13.** Problem 32.

I and II are a half-orbit apart. (a) Show that the star would appear to go from position (I) to position (II) in a time $T/2 - 2lv/(c^2 - v^2)$ and from position (II) to position (I) in a time $T/2 + 2lv/(c^2 - v^2)$, assuming that the emission theories are correct. (b) Show that the star would appear to be at both positions I and II at the same time if $T/2 = 2lv/(c^2 - v^2)$.

**33. Three emission theories (all wrong!)** Emission theories differ in their predictions of what the speed of light will be on reflection from a moving mirror. (*a*) The *original-source* theory assumes that the speed remains $c$ relative to the source. (*b*) The *ballistic* theory assumes that the speed becomes $c$ relative to the mirror. (*c*) The *new-source* theory assumes that the speed becomes $c$ relative to the mirror image of the source. Figure 1-14 shows a source of light $S$ moving to the left with speed $u$ in the laboratory frame and a mirror $M$ moving to the right with speed $v$. What is the speed of the reflected light beam according to each of the above three theories? (Compare Problem 34.)

**34. Moving sources, moving mirrors, and Einstein's second postulate.** According to Einstein's second postulate, what would

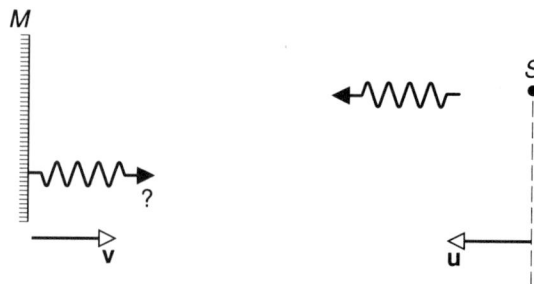

**FIGURE 1-14.**    Problems 33 and 34.

be the measured speed of a light pulse (both before and after reflection from the moving mirror of Fig. 1-14) as viewed from (*a*) the laboratory reference frame, (*b*) a frame attached to the source, and (*c*) a frame attached to the mirror? Compare the simplicity of your answers with those predicted by the various emission theories described in Problem 33.

# REFERENCES

1. ROBERT RESNICK AND DAVID HALLIDAY, *Physics,* Part I, 3rd ed. (Wiley, New York, 1977) and David Halliday and Robert Resnick, *Physics,* Part II, 3rd ed. (Wiley, New York, 1978).
2. WOLFGANG K. H. PANOFSKY AND MELBA PHILLIPS, *Classical Electricity and Magnetism,* 2nd ed., (Addison–Wesley, Reading, Mass., 1962), chap. 14.
3. R. S. SHANKLAND, "The Michelson-Morley Experiment," *Am. J. Phys.* **32,** 16 (1964). A detailed account of this experiment and a clear portrayal of the centrality of the ether concept in prerelativity days. Also, a long extract from the Michelson-Morley paper is given in W. F. Magie, *Source Book in Physics* (McGraw-Hill, New York, 1935), p. 369.
4. J. P. CEDARHOLM, G. L. BLAND, B. L. HAVENS, AND C. H. TOWNES, "New Experimental Tests of Special Relativity," *Phys. Rev. Lett.,* **1,** 342 (1958).
5. SIR EDMUND WHITTAKER, "G. F. Fitzgerald," *Scientific American* (November 1953). An account of Fitzgerald's life and the contraction hypothesis.
6. ALBERT STEWART, "The Discovery of Stellar Aberration," *Scientific American* (March 1964). An interesting and detailed description of Bradley's work.
7. J. G. FOX, "Evidence Against Emission Theories," *Am. J. Phys.,* **33,** 1 (1965). A critical analysis of the de Sitter and other experiments. See also J. G. Fox, "Constancy of the Velocity of Light," *Am. J. Phys.,* **35,** 967 (1967).
8. KENNETH BRECHER, "Is the Speed of Light Independent of the Velocity of the Source?" *Phys. Rev. Lett.,* **39,** 1051 (1977).
9. See, for example, T. ALVAGER, F. J. M. FARLEY, J. KJELLMAN, AND I. WALLEN, "Test of the Second Postulate of Special Relativity in the GeV Region," *Phys. Lett.,* **12,** 260 (1964).
10. A full translation of this 1905 paper by Einstein, along with translations of other significant papers by Einstein and by others, appears in *The Principle of Relativity* (Dover, New York, 1952).
11. GERALD HOLTON, "On the Origins of the Special Theory of Relativity," *Am. J. Phys.* **28,** 627 (1960).
12. R. S. SHANKLAND, "Conversations with Einstein," *Am. J. Phys.,* **31,** 47 (1963).
13. MARTIN J. KLEIN, "Einstein and Some Civilized Discontents," *Phys. Today,* **18,** 38 (1965).
14. GERALD HOLTON, "Influences on Einstein's Early Work," *Am. Scholar* (Winter 1967–68).

15. GERALD HOLTON, "Einstein and the 'Crucial' Experiment," *Am. J. Phys.,* **37,** 968 (1969).

16. JEREMY BERNSTEIN, *Einstein* (Viking, New York, 1973).

17. A. P. FRENCH, ED., *Einstein, A Centenary Volume,* (Harvard University Press, Cambridge, Mass., 1979). Authorized by the International Commission on Physics Education in celebration of the centenary of Einstein's birth.

18. ROBERT RESNICK, "Misconceptions About Einstein, His Work and His Views," *J. Chem. Educ.,* **57,** 854 (1980).

19. ABRAHAM PAIS, *'Subtle is the Lord . . . '—The Science and Life of Albert Einstein* (Oxford University Press, New York, 1982).

## CHAPTER 2

# Relativistic Kinematics

*We will raise this conjecture (whose intent will from now on be referred to as the "Principle of Relativity") to the status of a postulate, and also introduce another postulate, which is only apparently irreconcilable with the former: light is always propagated in empty space with a definite velocity* c *which is independent of the state of motion of the emitting body.*

*Albert Einstein (1905)*

## 2-1 The Relativity of Simultaneity

In a paper entitled "Conversations with Albert Einstein," R. S. Shankland* writes: "I asked Professor Einstein how long he had worked on the Special Theory of Relativity before 1905. He told me that he had started at age 16 and worked for ten years; first as a student when, of course, he could spend only part-time on it, but the problem was always with him. He abandoned many fruitless attempts, 'until at last it came to me that time was suspect!' " What was it about time that Einstein questioned? It was the assumption, often made unconsciously and certainly not stressed, that there exists a universal time that is the same for all observers. Indeed, it was only to bring out this assumption explicitly that we included the equation $t = t'$ in the Galilean transformation equations (Eq. 1-1). In prerelativistic discussions, the assumption was there implicitly by the absence of a transformation equation for $t$ in the Galilean equations. That the same time scale applied to all inertial frames of reference was a basic premise of Newtonian mechanics.

In order to set up a universal time scale, we must be able to give meaning, independent of a frame of reference, to statements such as "Events $A$ and $B$ occurred at the same time." Einstein pointed out that when we say that a train arrived at 7 o'clock, this means that the exact pointing of the clock hand to 7 and the arrival of the train at the clock were simultaneous. We certainly shall not have a universal time scale if different inertial observers disagree as to whether two events are simultaneous. Let us first try to set up an unambiguous time scale in a single frame of reference; then we can set up time scales in exactly the same way in all inertial frames and compare what different observers have to say about the sequence of two events, $A$ and $B$.

*See Ref. 12 in Chapter 1.

Suppose that the events occur at the same place in one particular frame of reference. We can have a clock at that place which registers the time of occurrence of each event. If the reading is the same for each event, we can logically regard the events as simultaneous. But what if the two events occur at *different* locations? Imagine now that there is a clock at the position of each event—the clock at $A$ being of the same nature as that at $B$, of course. These clocks can record the time of occurrence of the events but, before we can compare their readings, we must be sure that they are synchronized.

Some "obvious" methods of synchronizing clocks turn out to be erroneous. For example, we can set the two clocks so that they always read the same time as seen by the observer at the site of event $A$ (observer $A$). This means that whenever $A$ looks at the $B$ clock, it reads the same to him as his clock. The defect here is that if observer $B$ uses the same criterion (that is, that the clocks are synchronized if they always read the same time to *him*), he will find that the clocks are *not* synchronized if $A$ says that they *are*. The reason is that this method neglects the fact that it takes time for light to travel from $B$ to $A$ and vice versa. The student should be able to show that, if the distance between the clocks is $L$, one observer will see the other clock lag his by $2L/c$ when the other observer claims that they are synchronous. We certainly cannot have observers in the same reference frame disagree on whether clocks are synchronized or not, so we reject this method.

An apparent way out of this difficulty is simply to set the two clocks to read the same time and then move them to the positions where the events occur. (In principle, we need clocks everywhere in our reference frame to record the time of occurrence of events, but once we know how to synchronize two clocks we can, one by one, synchronize all the clocks.) The difficulty here is that we do not know ahead of time, and therefore cannot assume, that the motion of the clocks (which may have different velocities, accelerations, and path lengths in being moved into position) will not affect their readings or time-keeping ability. Even in classical physics, the motion can affect the rate at which clocks run. For example, a pendulum clock in free fall will not run at all!

Hence, the logical thing to do is to put our clocks into position and synchronize them by means of signals. If we had a method of transmitting signals with infinite speed, there would be no complications. The signals would go from clock $A$ to clock $B$ to clock $C$, and so on, in zero time. We could use such a signal to set all clocks at the same time reading. But no signal known has this property. All known signals require a finite time to travel some distance, the time increasing with the distance traveled. The best signal to choose would be one whose speed depends on as few factors as possible. We choose electromagnetic waves because they do not require a material medium for transmission and their speed in vacuum does not depend on their wavelength, amplitude, or direction of propagation. Furthermore, their propagation speed is the highest known and—most important for finding a universal method of synchronization—experiment shows their speed to be the same for all inertial observers.

Now we must account for the finite time of transmission of the signal and our clocks can be synchronized. To do this let us imagine an observer with a light source that can be turned on and off (for example, a flashbulb) at each clock, $A$ and $B$. Let the measured distance between the clocks (and observers) be $L$. The agreed-upon procedure for synchronization then is that $A$ will turn on his light source when his clock reads $t = 0$ and observer $B$ will set his clock to $t = L/c$ the instant he receives the signal. This accounts for the transmission time and synchronizes the clocks in a consistent way. For example, if $B$ turns on his light source at some later time $t$ by his clock, the signal will arrive at $A$ at a time $t + L/c$, which is just what $A$'s clock will read when $A$ receives the signal.

A method equivalent to the above is to put a light source at the exact midpoint of the straight line connecting $A$ and $B$ and instruct each observer to put his clock at $t = 0$ when the turned-on light signal reaches him. The light will take an equal amount of time to reach $A$ and $B$ from the midpoint, so this procedure does indeed synchronize the clocks.

Now that we have a procedure for synchronizing clocks in one reference frame, we can judge the time order of events in that frame. The time of an event is measured by the clock whose location coincides with that of the event. Events occurring at two different places in that frame must be called *simultaneous* when the clocks at the respective places record the same time for them. Suppose that one inertial observer does find that two separated events are simultaneous. Will these same events be measured as simultaneous by an observer on another inertial frame that is moving with speed $v$ with respect to the first? (Remember, each observer uses an identical procedure to synchronize the clocks in his reference frame.) If not, simultaneity is not independent of the frame of reference used to describe events. Instead of being absolute, simultaneity would be a relative concept. Indeed, this is exactly what we find to be true, in direct contradiction to the classical assumption.

We can understand the relativity of simultaneity by a simple example in which, in fact, clocks play no role. It will also be clear from this example in what a direct and compelling way the relativity of simultaneity follows from Einstein's second postulate, namely, that the speed of light is the same for all inertial observers. Let there be two inertial reference frames $S'$ and $S$, having a relative velocity $v$. Each observer is provided with a measuring rod. The observers note that two lightning bolts strike each frame, hitting and leaving permanent marks.* Assume that afterwards, by using their measuring rods, each inertial observer finds that he was located exactly at the midpoint of the marks that were left on his reference frame. In Fig. 2-1$a$, these marks are left at $A$ and $B$ on the $S$ frame and at $A'$ and $B'$ on the $S'$ frame, the observers being at $O$ and $O'$. Because each observer knows he was at the midpoint of the marks left by these events, he will conclude that the events were simultaneous if the light signals from them arrive simultaneously at his position (see the definitions of simultaneity given earlier). If, on the other hand, one signal arrives before the other, he will conclude that one event preceded the other.

Many different possibilities exist in principle as to what the measurements might show. Let us suppose, for the sake of argument, that the $S$ observer finds that the lightning bolts struck simultaneously. Will the $S'$ observer also find these events to be simultaneous? In Fig. 2-1 we take the point of view of the $S$ observer and see the $S'$ frame moving, say, to the right. At the instant the lightning struck at $A$ and $A'$,

*The essential point is to have light sources that leave marks. Note that, because the speed of light is the same in all inertial frames, only a single expanding wavefront originates at each pair of marks.

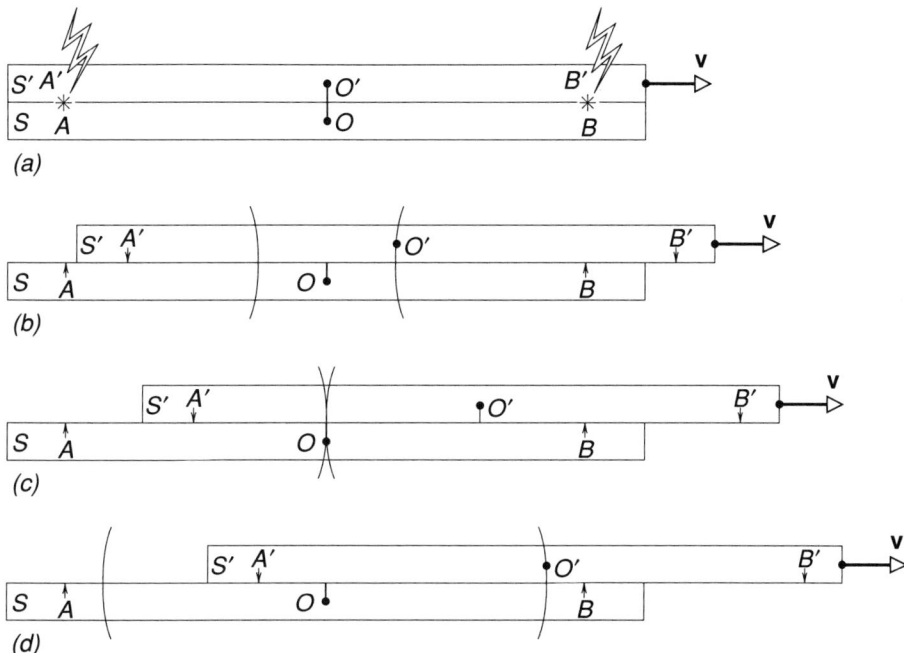

**FIGURE 2·1.** *The point of view of the S observer,* the $S'$ frame moving to the right. (a) Light waves leave $AA'$ and $BB'$. Successive drawings correspond to the assumption that event $AA'$ and event $BB'$ are simultaneous in the $S$ frame. (b) The right wavefront reaches $O'$. (c) Both wavefronts reach $O$. (d) The left wavefront reaches $O'$.

these two points coincide, and at the instant the lightning struck at $B$ and $B'$ those two points coincide. The $S$ observer found these two events to occur at the same instant, so at that instant $O$ and $O'$ must coincide also for him. However, *the light signals from the events take a finite time to reach $O$ and during this time $O'$ travels to the right* (Figs. 2-1$b$ to 2-1$d$). Hence, the signal from event $BB'$ arrives at $O'$ (Fig. 2-1$b$) before it gets to $O$ (Fig. 2-1$c$), whereas the signal from event $AA'$ arrives at $O$ (Fig. 2-1$c$) before it gets to $O'$ (Fig. 2-1$d$). Consistent with our starting assumption, the $S$ observer finds the events to be simultaneous (both signals arrive at $O$ at the same instant). The $S'$ observer, however, finds that event $BB'$ precedes event $AA'$ in time; they are *not* simultaneous to him. Since $O'$ is to the right of $O$ at all times after the strokes occur, the right wavefront *must* pass $O'$ before it reaches $O$. Thus, the two wavefronts *cannot* reach $O'$ simultaneously, and observer $S'$ *must* judge the two events to be nonsimultaneous.

Thus we see that two separated events which are judged to be simultaneous in one reference frame will, in general, not be so judged by an observer in a different frame. Note in what a direct and simple way this concept of the relativity of simultaneity follows from the principle of the constancy of the speed of light. If the speed of light were not the same for the two observers in Fig. 2-1, the entire argument we have just given would fall apart.

Now we could have supposed, just as well, that the lightning bolts struck so that the $S'$ observer found them to be simultaneous. In that case the light signals reach

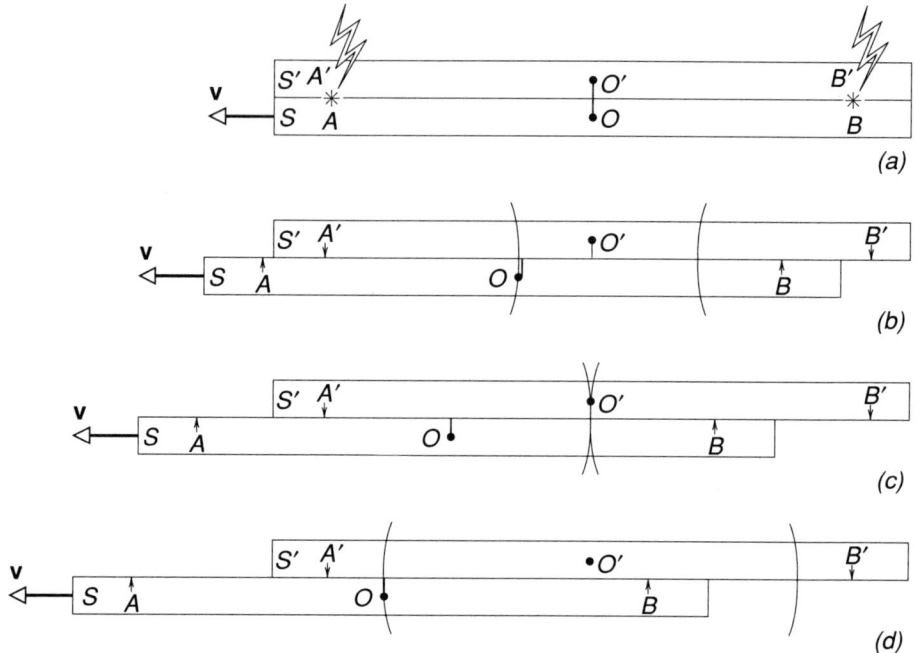

**FIGURE 2-2.** *The point of view of the S' observer,* the S frame moving to the left. (a) Light waves leave AA' and BB'. Successive drawings correspond to the assumption that event AA' and event BB' are simultaneous in the S' frame. (b) The left wavefront reaches O. (c) Both wavefronts reach O'. (d) The right wavefront reaches O.

O' simultaneously, rather than O. We show this in Fig. 2-2, where now we take the point of view of S'. The S frame moves to the left relative to the S' observer. But, in this case, the signals do not reach O simultaneously; the signal from event AA' reaches O before that from event BB'. Here the S' observer finds the events to be simultaneous but the S observer finds that event AA' precedes event BB'.

Hence, *neither* frame is preferred and the situation is perfectly reciprocal. Simultaneity is genuinely a relative concept, not an absolute one. Indeed, the two figures become indistinguishable if you turn one of them upside down. Neither observer can assert absolutely that he is at rest. Instead, each observer correctly states only that the other one is moving relative to him and that the signals travel with (the same) finite speed c relative to him. It should be clear that if we had an infinitely fast signal, then simultaneity *would* be an absolute concept; for the frames would not move at all relative to one another in the (zero) time it would take the signal to reach the observers.

Note that the time order of two events *at the same place* can be absolutely determined. It is in the case that two events are *separated* in space that simultaneity is a relative concept. In our arguments, we have shown that if one observer finds the events to be simultaneous, *then* the other one will find them not to be simultaneous. Of course, it could also happen that neither observer finds the events to be simultaneous, but then they would disagree either on the time order of the events or on the time interval elapsing between the events, or both.

Some other conclusions suggest themselves from the relativity of simultaneity. To measure the length of an object (a rod, say) means to locate its end points simultaneously. Because simultaneity is a relative concept, length measurements will also depend on the reference frame and be relative. Furthermore, we find that the rates at which clocks run also depend on the reference frame. We can illustrate this as follows. Consider two clocks, one on a train and one on the ground, and assume that at the moment they pass one another (that is, the instant that they are coincident), they read the same time (that is, the hands of the clocks are in identical positions). Now, if the clocks continue to agree, we can say that they go at the same rate. But, when they are a great distance apart, we know from the preceding discussion that their hands cannot have identical positions simultaneously as measured both by the ground observer and the train observer. Hence, time interval measurements are also relative; that is, they depend on the reference frame of the observer.

## 2-2 Derivation of the Lorentz Transformation Equations

We have seen that the Galilean transformation equations must be replaced by new ones consistent with experiment. Here we shall derive these new equations, using the postulates of special relativity theory. To show the consistency of the theory with the discussion of the previous section, we shall then derive all the special features of the new transformation equations again from the more physical approach of the measurement processes discussed there.

We observe an event in one inertial reference frame $S$ and characterize its location and time by recording the coordinates $x, y, z, t$ of the event. In a second inertial frame $S'$, this *same event* is recorded by the observer there as the space-time coordinates $x', y', z', t'$. We now seek the functional relationships $x' = x'(x, y, z, t)$, $y' = y'(x, y, z, t)$, $z' = z'(x, y, z, t)$, and $t' = t'(x, y, z, t)$. That is, we want the equations of transformation which relate one observer's space-time coordinates of an event with the other observer's coordinates of the same event.

We shall use the fundamental postulates of relativity theory and, in addition, the assumption that space and time are homogeneous. This homogeneity assumption (which can be paraphrased by saying that all points in space and time are equivalent) means, for example, that the results of a measurement of a length or time interval between two specific events should not depend on where or when the interval happens to be in our reference frame. We shall illustrate its application shortly.

We can simplify the algebra by choosing the relative velocity of the $S$ and $S'$ frames to be along a common $x$-$x'$ axis and by keeping corresponding planes parallel (See Fig. 1-1). This does not impose any fundamental restrictions on our results, for space is isotropic (that is, has the same properties in all directions), a result contained in the homogeneity assumption. Also, at the instant the origins $O$ and $O'$ coincide, we let the clocks there read $t = 0$ and $t' = 0$, respectively. Now, as explained below, the homogeneity assumption requires that transformation equations must be linear (that is, they involve only the first power in the variables), so that the most general form they can take is

$$x' = a_{11}x + a_{12}y + a_{13}z + a_{14}t$$
$$y' = a_{21}x + a_{22}y + a_{23}z + a_{24}t$$
$$z' = a_{31}x + a_{32}y + a_{33}z + a_{34}t$$
$$t' = a_{41}x + a_{42}y + a_{43}z + a_{44}t$$

(2-1)

Here, the subscripted coefficients are constants that we must determine to obtain the exact transformation equations. Notice that we do not exclude the possible dependence of space and time coordinates on one another.

If the equations were not linear, we would violate the homogeneity assumption. For example, suppose that $x'$ depended on the square of $x$, that is, as $x' = a_{11}x^2$. Then the distance between two points in the primed frame would be related to the location of these points in the unprimed frame by $x'_2 - x'_1 = a_{11}(x_2^2 - x_1^2)$. Suppose now that a rod of unit length in $S$ had its end points at $x_2 = 2$ and $x_1 = 1$; then $x'_2 - x'_1 = 3a_{11}$. If, instead, the same rod happens to be located at $x_2 = 5$ and $x_1 = 4$, we would obtain $x'_2 - x'_1 = 9a_{11}$. That is, the measured length of the rod would depend on where it is in space. Likewise, we can reject any dependence on $t$ that is not linear, for the time interval between two events should not depend on the numerical setting of the hands of the observer's clock. The relationships must be linear then in order not to give the choice of origin of our space-time coordinates (or some other point) a physical preference over all other points.

Now, regarding the 16 coefficients in Eq. 2-1, it is expected that their values will depend on the relative velocity $v$ of the two inertial frames. For example, if $v = 0$, then the two frames coincide at all times and we expect $a_{11} = a_{22} = a_{33} = a_{44} = 1$, all other coefficients being zero. More generally, if $v$ is small compared to $c$, the coefficients should lead to the (classical) Galilean transformation equations. We seek to find the coefficients for *any* value of $v$, that is, as functions of $v$.

How then do we determine the values of these 16 coefficients? Basically, we use the postulates of relativity, namely, (1) the principle of relativity—that no preferred inertial system exists, the laws of physics being the same in all inertial systems—and (2) the principle of the constancy of the speed of light—that the speed of light in free space has the same value $c$ in all inertial systems. Let us proceed.

With no relative motion of the frames in the $y$ or $z$ direction, we might expect $y' = y$ and $z' = z$. This result does indeed follow directly from arguments using the relativity postulate (see Sec. 2-2 of Ref. 1 for a proof), so that $a_{22} = a_{33} = 1$ and $a_{21} = a_{23} = a_{24} = a_{31} = a_{32} = a_{34} = 0$ and eight of the coefficients are thereby determined. Therefore, our two middle transformation equations become

$$y' = y \quad \text{and} \quad z' = z.$$

(2-2)

There remain transformation equations for $x'$ and $t'$, namely,

$$x' = a_{11}x + a_{12}y + a_{13}z + a_{14}t$$

and

$$t' = a_{41}x + a_{42}y + a_{43}z + a_{44}t.$$

Let us look first at the $t'$ equation. For reasons of symmetry, we assume that $t'$ does not depend on $y$ and $z$. Otherwise, clocks placed symmetrically in the $y$-$z$ plane (such as at $+y$, $-y$ or $+z$, $-z$) about the $x$ axis would appear to disagree as observed for $S'$, which could contradict the isotropy of space. Hence, $a_{42} = a_{43} = 0$. As for the $x'$ equation, we know that a point having $x' = 0$ appears to move in the direction of the positive $x$ axis with speed $v$, so that the statement $x' = 0$ must be identical to the statement $x = vt$. Putting $x' = 0$ and $x = vt$ in the $x'$ equation above then yields

$$0 = a_{11}(vt) + a_{12}y + a_{13}z + a_{14}t$$

or

$$0 = (a_{11}v + a_{14})t + a_{12}y + a_{13}z.$$

Because $t$, $y$, and $z$ are all independent variables, the only way this last equation can be satisfied for all values of these variables is for the coefficients of the variables to vanish. Thus we must have

$$a_{14} = -a_{11}v$$

and

$$a_{12} = a_{13} = 0.$$

Our four transformation equations are thus reduced to

$$\begin{aligned} x' &= a_{11}(x - vt) \\ y' &= y \\ z' &= z \\ t' &= a_{41}x + a_{44}t. \end{aligned} \tag{2-3}$$

There remains the task of determining the three coefficients $a_{11}$, $a_{41}$, and $a_{44}$. To do this, we use the principle of the constancy of the velocity of light. Let us assume that at the time $t = 0$ a spherical electromagnetic wave leaves the origin of $S$, which coincides with the origin of $S'$ at that moment. The wave propagates with a speed $c$ in all directions in each inertial frame. Its progress, then, is described by the equation of a sphere whose radius expands with time at the same rate $c$ in terms of *either* the primed or unprimed set of coordinates. That is,

$$x^2 + y^2 + z^2 = c^2t^2 \tag{2-4}$$

or

$$x'^2 + y'^2 + z'^2 = c^2t'^2. \tag{2-5}$$

If now we substitute the transformation equations (Eqs. 2-3) into Eq. 2-5, we get

$$a_{11}^2(x - vt)^2 + y^2 + z^2 = c^2(a_{41}x + a_{44}t)^2.$$

Rearranging the terms gives us

$$(a_{11}^2 - c^2a_{41}^2)x^2 + y^2 + z^2 - 2(va_{11}^2 + c^2a_{41}a_{44})xt = (c^2a_{44}^2 - v^2a_{11}^2)t^2.$$

In order for this expression to agree with Eq. 2-4, which represents the same thing, we must have

$$a_{11}^2 - c^2a_{41}^2 = 1,$$
$$va_{11}^2 + c^2a_{41}a_{44} = 0,$$

and

$$c^2a_{44}^2 - v^2a_{11}^2 = c^2.$$

Here we have three equations in three unknowns, whose solution (as you can verify by substitution into the three equations above) is

$$a_{11} = \frac{1}{\sqrt{1 - v^2/c^2}},$$

$$a_{41} = -\frac{(v/c^2)}{\sqrt{1 - v^2/c^2}}, \qquad (2\text{-}6)$$

and

$$a_{44} = \frac{1}{\sqrt{1 - v^2/c^2}}.$$

By substituting these values into Eqs. 2-3, we obtain, finally, the new sought-after transformation equations,

$$x' = \frac{x - vt}{\sqrt{1 - v^2/c^2}}$$

$$y' = y$$

$$z' = z \qquad (2\text{-}7)$$

$$t' = \frac{t - (v/c^2)x}{\sqrt{1 - v^2/c^2}},$$

the so-called* *Lorentz transformation equations.*

Simple inspection of Eqs. 2-7 shows at once that serious difficulties arise if the relative speed $v$ of the two observers is equal to or greater than the speed of light. Under these circumstances the predicted quantities $x'$ and $t'$ are no longer finite real numbers. They are either infinitely great (for $v = c$) or imaginary† (for $v > c$). We conclude that $c$ represents a limiting speed and that the speeds of material objects must always be less than this value, regardless of the reference frame from which the objects are observed. This conclusion is in complete agreement with experiment, no exceptions ever having been found.

Before probing the meaning of the Lorentz equations, we should put them to two necessary tests. First, if we were to exchange our frames of reference or—what amounts to the same thing—consider the given space-time coordinates of the event to be those observed in $S'$ rather than in $S$, the only change allowed by the relativity principle is the physical one of a change in relative velocity from $v$ to $-v$ . That is, from $S'$ the $S$ frame moves to the left, whereas from $S$ the $S'$ frame moves to the right. When we solve Eqs. 2-7 for $x, y, z$, and $t$ in terms of the primed coordinates (see Problem 5), we obtain

$$x = \frac{x' + vt'}{\sqrt{1 - v^2/c^2}},$$

$$y = y',$$

$$z = z', \qquad (2\text{-}8)$$

---

*Poincaré originally gave this name to the equations. Lorentz, in his classical theory of electrons, had proposed them before Einstein did. However, Lorentz took $v$ to be the speed relative to an absolute ether frame and gave a different interpretation to the equations.

†In the mathematical sense, meaning a quantity containing $\sqrt{-1}$ as a factor.

$$t = \frac{t' + (v/c^2)x'}{\sqrt{1 - v^2/c^2}},$$

which are identical in form with Eqs. 2-7 except that, as required, $v$ changes to $-v$.

Another requirement is that for speeds small compared to $c$, that is, for $v/c \ll 1$, the Lorentz equations should reduce to the (approximately) correct Galilean transformation equations. A formal but useful way of imposing the requirement that the speed $v$ is very much less than $c$, the speed of light, is to imagine a world in which the speed of light is infinitely great. In such a world, $v/c \ll 1$ would *always* be true, for all (finite) speeds. If, then, we let $c \rightarrow \infty$ in Eqs. 2-7, we obtain

$$\begin{aligned} x' &= x - vt \\ y' &= y \\ z' &= z \\ t' &= t, \end{aligned} \tag{2-9}$$

which are the classical Galilean transformation equations.

In dealing algebraically with the Lorentz equations, and with other equations that we shall encounter later, it simplifies matters greatly to introduce two new parameters, $\beta$ and $\gamma$. Both are dimensionless, and both are simple functions of the relative speed $v$. The *speed parameter* $\beta\ (= v/c)$ is simply the ratio of the relative speed $v$ to the speed of light. The second parameter, $\gamma$, often called the *Lorentz factor*, is defined from

$$\gamma = \frac{1}{\sqrt{1 - v^2/c^2}} = \frac{1}{\sqrt{1 - \beta^2}}. \tag{2-10}$$

Table 2-1 shows some selected values of $\beta$ along with the corresponding values of $\gamma$ computed from Eq. 2-10. Figure 2-3 is a plot of Eq. 2-10. In the classical low-speed limit, which is described by $\beta \ll 1$, we see that $\gamma \rightarrow 1$. In the high-speed limit, which is described by $\beta \rightarrow 1$, we see that $\gamma \rightarrow \infty$.

Table 2-2 displays the Lorentz transformation equations (Eqs. 2-7 and 2-8) with $\gamma$ written in place of the quantity $1/\sqrt{1 - v^2/c^2}$.

The Lorentz transformation equations displayed in Table 2-2 refer to a single event whose space-time coordinates are measured by two different inertial observers, $S$ and $S'$. Commonly, however, we are confronted with pairs of events (event 1 and event 2, say) and our interest centers on the differences between their corresponding space-time coordinates rather than on the coordinates themselves. We

**TABLE 2-1**
Some Values of $\gamma$

| $\beta$ | $\sqrt{1-\beta^2}$ | $\gamma$ | $\beta$ | $\sqrt{1-\beta^2}$ | $\gamma$ |
|---|---|---|---|---|---|
| 0.000 | 1.000 | 1.000 | 0.900 | 0.437 | 2.29 |
| 0.050 | 0.9987 | 1.0013 | 0.950 | 0.312 | 3.21 |
| 0.100 | 0.9950 | 1.0050 | 0.990 | 0.141 | 7.09 |
| 0.300 | 0.9542 | 1.048 | 0.9990 | 0.0446 | 22.4 |
| 0.600 | 0.8000 | 1.25 | 0.99990 | 0.0141 | 70.7 |

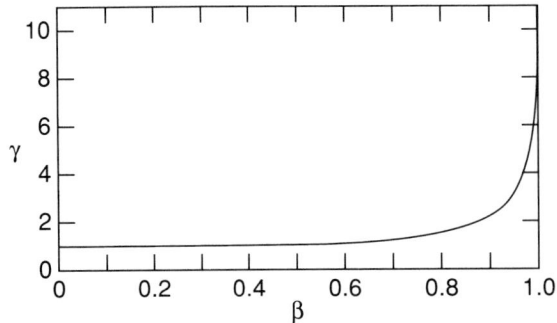

**FIGURE 2-3.** A plot of $\gamma$ $(= 1/\sqrt{1-\beta^2})$ against $\beta$ $(= v/c)$. See Eq. 2-10 and Table 2-1.

**TABLE 2-2**
The Lorentz Transformation Equations

| | |
|---|---|
| 1. $x = \gamma(x' + vt')$ | 1'. $x' = \gamma(x - vt)$ |
| 2. $y = y'$ | 2'. $y' = y$ |
| 3. $z = z'$ | 3'. $z' = z$ |
| 4. $t = \gamma\left(t' + \dfrac{vx'}{c^2}\right)$ | 4'. $t' = \gamma\left(t - \dfrac{vx}{c^2}\right)$ |

$$\gamma = \frac{1}{\sqrt{1 - v^2/c^2}}$$

are often more concerned, for example, with $\Delta x$ $(= x_2 - x_1)$ and with $\Delta t$ $(= t_2 - t_1)$ than with the separate quantities $x_2$, $x_1$, $t_2$, and $t_1$. Table 2-3 shows the Lorentz transformation equations written in a difference form, suitable to such occasions.

**TABLE 2-3**
The Lorentz Transformation Equations[a]

| | |
|---|---|
| 1. $\Delta x = \gamma(\Delta x' + v\,\Delta t')$ | 1'. $\Delta x' = \gamma(\Delta x - v\,\Delta t)$ |
| 2. $\Delta y = \Delta y'$ | 2'. $\Delta y' = \Delta y$ |
| 3. $\Delta z = \Delta z'$ | 3'. $\Delta z' = \Delta z$ |
| 4. $\Delta t = \gamma\left(\Delta t' + \dfrac{v\,\Delta x'}{c^2}\right)$ | 4'. $\Delta t' = \gamma\left(\Delta t - \dfrac{v\,\Delta x}{c^2}\right)$ |

[a] Written for event pairs, as difference equations.

## EXAMPLE 1.

**Two Observers View a Single Event.** An observer in the $S$ frame of Fig. 1-1 notes that an event occurs along the $x$-$x'$ axis and records its space-time coordinates as $x = 2.00$ m and $t = 5.00$ ns $(= 5.00 \times 10^{-9}$ s). The $S'$ frame is moving with a speed $v$ with respect to the $S$ frame along their common axis. (a) If $v = 0.500c$, what space-time coordinates would the $S'$ observer record for this event? Take the speed of light $c$ to be $3.00 \times 10^8$ m/s, noting that this can be usefully written as 0.300 m/ns. (*b*) What space-time coordinates would observer $S'$ record if the Galilean transformation equations held?

(*a*) For $\beta = 0.500$ we have, from Eq. 2-10,
$$\gamma = \frac{1}{\sqrt{1 - \beta^2}} = \frac{1}{\sqrt{1 - (0.50)^2}} = 1.16,$$

and, from Table 2-2 (Eqs. 1' and 4'),
$$x' = \gamma(x - vt)$$
$$= (1.16)[2.00 \text{ m} - (0.500 \times 0.300 \text{ m/ns})$$
$$(5.00 \text{ ns})]$$
$$= (1.16)(1.25 \text{ m}) = 1.45 \text{ m},$$

and
$$t' = \gamma\left(t - \frac{vx}{c^2}\right)$$
$$= \gamma\left(t - \frac{\beta x}{c}\right)$$
$$= (1.16)\left[5.00 \text{ ns} - \frac{(0.500)(2.00 \text{ m})}{(0.300 \text{ m/ns})}\right]$$
$$= (1.16)(1.67 \text{ ns}) = 1.94 \text{ ns}.$$

(*b*) The Galilean transformation equations (Eq. 2-9) yield
$$x' = x - vt$$

$$= 2.00 \text{ m} - (0.500 \times 0.300 \text{ m/ns})(5.00 \text{ ns})$$
$$= 1.25 \text{ m}$$
and
$$t' = t = 5.00 \text{ ns}.$$

Summarizing the results for easy comparison:

| Observer S | Observer S' | |
|---|---|---|
| | Lorentz Equations | Galilean Equations |
| $x$  2.00 m | $x'$  1.45 m | (1.25 m) |
| $t$  5.00 ns | $t'$  1.94 ns | (5.00 ns) |

We see that for such a large relative speed (half the speed of light!), the Lorentz transformation equations, which are correct for all speeds, predict results that differ substantially from the predictions of the Galilean equations. The Galilean predictions do not agree with the observations, and we have enclosed them in parentheses to remind us of that fact.

Recall that the length measurements ($x$ and $x'$) are from the appropriate origin ($O$ or $O'$) to the event. The time measurements ($t$ and $t'$) are the times that have elapsed since the moment the origins of the two reference frames passed each other. Each observer set his clocks to zero at that instant.

---

# EXAMPLE 2.

***The Galilean Equations Fail at High Speeds.***
At what relative speed will the Galilean and the Lorentz expressions for position $x$ differ by 0.10 percent? By 1.0 percent? By 10 percent?

We can write the Galilean expression as $x'_G = x - vt$ and the Lorentz expression (see Table 2-2; Eq. 1') as $x'_L = \gamma(x - vt)$. We seek the value of $v$ for which
$$\frac{x'_L - x'_G}{x'_L} = 0.10\% = 0.001.$$

We can write this as
$$\frac{x'_G}{x'_L} = 1 - 0.001 = 0.9990.$$

Substituting from above yields
$$\frac{(x - vt)}{\gamma(x - vt)} = 0.9990$$
or
$$\frac{1}{\gamma} = 0.9990.$$

From Eq. 2-10, the defining equation for $\gamma$, we then have
$$\frac{1}{\gamma} = \sqrt{1 - \beta^2} = 0.9990.$$

Solving for $\beta$ yields
$$\beta = 0.045.$$

Thus

$$v = \beta c = (0.045)(3.00 \times 10^8 \text{ m/s})$$

$$= 1.4 \times 10^7 \text{ m/s}$$

is the answer we seek.

At this value of $v$ (almost 5 percent of the speed of light) the Lorentz and the Galilean transformation equations differ in their predictions by only 0.1 percent. Small as this difference might appear, it still occurs at a speed (32 million mi/h!) much greater than any speed we encounter in the macroscopic world. We are quite safe in using the Galilean transformation equations when dealing with the macroscopic objects such as baseballs or spaceships.

For a difference of 1 percent we find similarly that $\beta = 0.14$, or

$$v = \beta c = (0.14)(3.00 \times 10^8 \text{ m/s})$$

$$= 4.2 \times 10^7 \text{ m/s}.$$

For a difference of 10 percent we find $\beta = 0.44$ and $v = 1.3 \times 10^8$ m/s. This latter speed, which is 44 percent of the speed of light, may seem high—and indeed it is impossibly so for terrestrial macroscopic objects. However, as we shall see in later sections, it corresponds to an electron with a kinetic energy of only ~60 keV, a modest laboratory achievement. We shall encounter speeds much higher than this later in examples from microscopic physics. For example, electrons emerging from the 2-mile-long Stanford Linear Accelerator with a kinetic energy of 30 GeV ($= 3 \times 10^{10}$ eV) have $v = 0.99999999986c$. At these enormous energies the Galilean equations are hopelessly inadequate; the Lorentz equations, however, continue to give results that agree with experiment.

## 2-3  Some Consequences of the Lorentz Transformation Equations

The Lorentz transformation equations (Eqs. 2-7 and 2-8 and Table 2-2), derived rather formally in the last section from the relativity postulates, have some interesting consequences for length and time measurements. We shall look at them briefly in this section. In the next section we shall present a more physical interpretation of these equations and their consequences, relating them directly to the operations of physical measurement. Throughout the chapter we shall cite experiments that confirm these consequences.

**Moving Rods Contract**  The Lorentz transformation equations predict that: *When a body moves with a velocity v relative to the observer, its measured length is contracted in the direction of its motion by the factor* $\sqrt{1 - v^2/c^2}$, *whereas its dimensions perpendicular to the direction of motion are unaffected.* To prove the italicized statement, imagine a rod lying at rest along the $x'$ axis of the $S'$ frame. Its end points are measured to be at $x'_2$ and $x'_1$, so that its rest length is $x'_2 - x'_1$ ($= \Delta x'$). What is the rod's length as measured by the $S$-frame observer, for whom the rod moves with a relative speed $v$? From the Lorentz equations (Table 2-3, Eq. 1'), we have

$$\Delta x' = \gamma(\Delta x - v \, \Delta t). \tag{2-11}$$

We can identify $\Delta x$ as the length of the rod in the $S$ frame if (and only if) the positions $x_2$ and $x_1$ of the end points of the rod are measured in this frame *at the same time*. Thus, with $\Delta t = 0$ in the above we obtain

$$\Delta x' = \gamma \Delta x$$

or (see Eq. 2-10)

$$\Delta x = \Delta x'(1/\gamma) = \Delta x' \sqrt{1 - \beta^2}. \qquad (2\text{-}12a)$$

Putting $l$ ($= \Delta x$) for the length of the moving rod and $l_0$ ($= \Delta x'$) for its length at rest allows us to write

$$l = l_0 \sqrt{1 - \beta^2} \qquad \text{(length contraction)} \qquad (2\text{-}12b)$$

so that the measured length $l$ of the moving rod is contracted by the factor $\sqrt{1-\beta^2}$ from its rest length $l_0$. As for the dimensions of the rod along $y$ and $z$, perpendicular to the relative motion, it follows at once from the transformation equations $y' = y$ and $z' = z$ that these are measured to be the same by both observers.

Equation 2-12b predicts that a rod has its greatest length ($= l_0$) when it is at rest (that is, when $\beta = 0$). Finally, our conclusions about the contraction of a moving rod are completely symmetrical, as required by the principle of relativity. We have seen above that a rod fixed in the $S'$ frame is measured to be shorter by an observer in the $S$ frame. It is equally true that a rod at rest in the $S$ frame would be measured as shorter by an $S'$ observer.

**Moving Clocks Run Slow**  The Lorentz transformation equations also predict that *when a clock moves with a velocity $v$ with respect to an observer, its rate is measured to have slowed down by a factor* $\sqrt{1-(v/c)^2}$. To prove this, consider a clock $C$ to be at rest at the position $x'_0$ in the $S'$ frame, as shown in Figure 2-4a. *The figure shows two fiduciary marks on the clock face. Let $t'_1$ be the time at which the*

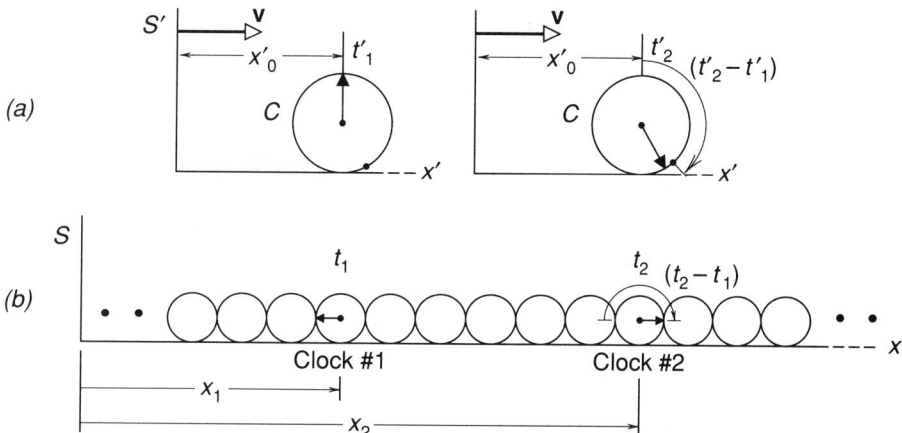

**FIGURE 2-4.**  Clock $C$ is fixed at position $x'_0$ in reference frame $S'$. Observer $S$ sees this as a *single moving clock* and compares its readings with *two different stationary clocks* from the array of synchronized clocks that he has established in his own reference frame. As inspection of the clock hands shows, the interval $t_2 - t_1$ is greater than the interval $t'_2 - t'_1$. Observer $S$ thus declares that the moving clock ($C$) is running slow by comparison with his own clocks.

hand of the clock, as seen by the $S'$ observer, coincides with the first of these marks, and let $t'_2$ be the time at which it coincides with the second.

The $S$-frame observer, on the other hand, records these same two events as occurring at times $t_1$ and $t_2$, as Fig. 2-4$b$ shows. It is important to realize that observer $S$ (like observer $S'$ for that matter) has at his disposal a whole array of synchronized clocks, imagined to be there just for the purpose of assigning time coordinates to various events. $S$ reads the times $t_1$ and $t_2$ from *two different* clocks in this array. These are the stationary $S$ clocks that happen to coincide in position with the moving clock $C$ both at the beginning and the end of the interval under consideration. The curved arrows in Fig. 2-4 identify the intervals $\Delta t'$ ($= t'_1 - t'_2$) and $\Delta t$ ($= t_1 - t_2$) as measured by the $S'$ observer and the $S$ observer, respectively, and show clearly that the first of these intervals is shorter than the second.

From the Lorentz transformation equations (Table 2-3, Eq. 4) we have

$$\Delta t = \gamma(\Delta t' + v\,\Delta x'/c^2). \qquad (2\text{-}13)$$

However, we have assumed that clock $C$ is at rest in the $S'$ frame, so that $\Delta x' = 0$. Putting this information into the above equation yields, for the time interval as measured by the $S$ observer, $\Delta t = \gamma\Delta t' = \Delta t'/\sqrt{1-\beta^2}$ which we can write as

$$t_2 - t_1 = \frac{t'_2 - t'_1}{\sqrt{1 - \beta^2}}. \qquad (2\text{-}14a)$$

Because $\sqrt{1-\beta^2} < 1$ we see that the interval $(t_2 - t_1)$ is greater than the interval $(t'_2 - t'_1)$, as reflected by the curved arrows on the clock faces in Fig. 2-4. A time interval measured on the single $S'$ clock is recorded as a *longer* time interval by the $S$ clocks. From the point of view of observer $S$, the moving $S'$ clock appears slowed down, that is, it appears to run at a rate which is slow by the factor $\sqrt{1-\beta^2}$. This result applies to all $S'$ clocks observed from $S$, for in our proof the location $x'_0$ was arbitrary.

In general, let us suppose that two events occur at a given place and let us represent by $\Delta\tau$ the time interval between them, measured on a clock at rest at that same place. Let $\Delta t$ be the interval between these same two events, measured by an observer for whom the single clock is moving. We can identify $\Delta\tau$ with $(t'_2 - t'_1)$ in Eq. 2-14$a$ and $\Delta t$ with $(t_2 - t_1)$ and write

$$\Delta t = \frac{\Delta\tau}{\sqrt{1 - \beta^2}}. \qquad \text{(time dilation)} \qquad (2\text{-}14b)$$

Equation 2-14$b$ predicts that a clock runs the fastest (that is, $\Delta\tau$ will equal $\Delta t$ rather than being smaller than $\Delta t$) when it is at rest (that is, when $\beta = 0$). Finally, our conclusions about clock rates are completely symmetrical, as required by the principle of relativity. We have seen that a clock fixed in the $S'$ frame appears to run slow as seen by an observer in the $S$ frame. It is equally true that a clock fixed in the $S$ frame would appear to run slow to an $S'$ observer.

It is common in relativity to speak of the frame in which the observed body is at rest as the *proper* frame. The length of a rod in such a frame is then called its *proper length* or, equivalently, its *rest length*. Likewise, the *proper time interval* is the time interval recorded by a clock attached to the observed body. The proper time interval can be thought of equivalently as the time interval between two events occurring at the same place in a given frame or the time interval measured by a single clock at one place. A nonproper time interval would be a time interval measured by two different clocks at two different places. Thus, in Eq. 2-14 (see also Fig. 2-4), the interval $\Delta\tau$ [$= (t'_2 - t'_1)$] is a proper interval, being recorded by a single clock ($C$) fixed in the $S'$ frame. The interval $\Delta t$ [$= (t_2 - t_1)$], on the other hand, is a nonproper interval, being recorded on two different clocks, separated by a distance $(x_2 - x_1)$ in the $S$ frame. Later we shall define other proper quantities, such as *proper mass* (often called *rest mass*), and shall examine the usefulness of these concepts in relativity theory.

**The Relativity of Clock Synchronization** A third consequence of the Lorentz transformation equations is this: Although clocks in a moving frame all appear to go at the same slow rate when observed from a stationary frame with respect to which the clocks move, *the moving clocks appear to differ from one another in their readings by a phase constant that depends on their location; that is, they appear to be unsynchronized*. This becomes evident at once from the transformation equation (see Table 2-2, Eq. 4)

$$t = \gamma \left( t' + \frac{vx'}{c^2} \right).$$

For consider an instant of time in the $S$ frame, that is, a given value of $t$. Then, to satisfy this equation, $t' + vx'/c^2$ must have a definite fixed value. This means the greater is $x'$ (that is, the farther away an $S'$ clock is stationed on the $x'$ axis), the smaller is $t'$ (that is, the farther behind in time its reading appears to be). Hence, the moving clocks appear to be out of phase, or synchronization, with one another. We shall see in the next section that this is just another manifestation of the fact that two events that occur simultaneously in the $S$ frame are not, in general, measured to be simultaneous in the $S'$ frame, and vice versa.

The lack of synchronization of clocks in a moving reference frame, like the contraction of moving rods and the slowing down of moving clocks, is also reciprocal. If observer $S$ declares that the clocks in her frame are synchronized but that those in the (moving) $S'$ frame are not, observer $S'$ can make a similar statement. He can, with equal validity, declare that the clocks in *his* reference frame are synchronized but that those in the (moving) $S$ frame are not.

**The Relativity of Simultaneity** In Section 2-1 we developed a physical argument (see Figs. 2-1 and 2-2) to show that observers in different reference frames cannot agree as to whether two events are simultaneous or not. Our argument there followed in a direct and logical way from the principle of the constancy of the speed of light. This concept of the relativity of simultaneity should also be contained in the Lorentz transformation equations, because they were derived from this same principle.

Consider two events, observed both by $S$ and by $S'$. The time differences between these events, as reported by these two observers, are related by Eq. 2-13, which follows directly from the Lorentz transformation equations (Table 2-3, Eq. 4). Thus,

$$(t_2 - t_1) = \gamma(t'_2 - t'_1) + \left( \frac{\gamma v}{c^2} \right) (x'_2 - x'_1). \tag{2-13}$$

We see at once that if $S'$ finds the events to be simultaneous (that is, if $t'_2 = t'_1$), then $S$ will *not* find them so, *unless* $S'$ also finds that the events occur in the same place, that is, unless $x'_2 = x'_1$. Note also that, if the events occur at different places it is possible for $S$ and $S'$ to disagree even about the sequence of the two events. If $t'_2 > t'_1$, for example, observer $S'$ declares that event 1 (having the smaller time of occurrence) comes first. However, if $x'_1$ is large enough (that is, if event 1 takes place far enough from the origin of the $S'$ frame), the last term in Eq. 2-13 can be sufficiently negative to require that $t_1 > t_2$. This means that observer $S$ will declare that event 2 comes first. All of this is in complete accord with what we have already learned in Section 2-1 by direct application of the principle of the constancy of the speed of light.

**The Spacetime\* Interval—An Invariant Quantity**   In the limit of low speeds, where the Galilean transformation equations hold sufficiently well for all practical purposes, the length of a rod is a fixed quantity, having the same value for all inertial observers. The same thing is true for the time interval between any two events. Quantities that have the same value for all inertial observers are said be *invariant*. We see that under a Galilean transformation there are at least two invariant quantities, one involving space coordinates and one the time coordinate.

Under a Lorentz transformation, however, we have seen that the lengths of rods and the time intervals between events are precisely *not* invariant. Different observers obtain, by measurement, different numerical results. The question naturally arises: Is there any quantity involving the space and time coordinates of events that *is* invariant under a Lorentz transformation? The answer proves to be "yes."

Consider two events, viewed by observers $S$ and $S'$. For simplicity, let us imagine that the events occur on the common $x$-$x'$ axis. We state without proof (see, however, Problem 41 and Supplementary Topic A) that the quantity

$$c^2(t_2 - t_1)^2 - (x_2 - x_1)^2,$$

which we can write as

$$(c \, \Delta t)^2 - (\Delta x)^2,$$

*is* invariant under a Lorentz transformation. By this we mean that although observer $S'$ would find, by measurement, that $\Delta t' \neq \Delta t$ and that $\Delta x' \neq \Delta x$, he would also find that

$$(c \, \Delta t')^2 - (\Delta x')^2 = (c \, \Delta t)^2 - (\Delta x)^2. \tag{2-15}$$

It is not unexpected that in relativity theory any quantity found to be invariant would involve both space *and* time coordinates. The invariance of the quantity displayed above has been described in a remarkable *tour de force* by W. A. Shurcliff entitled *Special Relativity—A back-of-an-envelope summary in words of one syllable;* see page 60.

The invariant quantity displayed in Eq. 2-15 is called the (square of the) *spacetime interval* (or, more commonly, simply the *interval*) between the two events and is symbolized by $(\Delta s)^2$. Thus,

---

\*To stress the intimate interconnection between space and time in relativity theory it is common to treat "spacetime" as a single word, without a hyphen.

$$(\Delta s)^2 = (c\ \Delta t)^2 - (\Delta x)^2. \tag{2-16}$$

For any given pair of events, $(\Delta s)^2$ is the only measureable characteristic kinematic quantity for which all inertial observers would obtain the same numerical value. Inspection of Eq. 2-16 shows that, although the two terms on the right of that equation are always positive, $(\Delta s)^2$ itself can be positive, negative, or zero, depending on the relative magnitudes of those terms. Note also that if $\Delta s^2$ is positive, say, in any given frame, it will be positive in all frames; it is an invariant quantiy.

For $(\Delta s)^2$ in Eq. 2-16 to be positive, the events must be such that $(c\ \Delta t)^2 > (\Delta x)^2$. It helps in visualizing such pairs of events to imagine them as spaced relatively close together along the $x$ axis ($\Delta x$ small) and/or separated by a relatively long time interval ($\Delta t$ large). The spacetime interval associated with such event pairs is described as *timelike* because the term containing $\Delta t$ predominates.

The *proper time* interval $\Delta \tau$ between any two events is defined as

$$\Delta \tau = \frac{\Delta s}{c} = \sqrt{(\Delta t)^2 - \left(\frac{\Delta x}{c}\right)^2}. \tag{2-17}$$

We see that $\Delta \tau$ is invariant because of the invariance of $\Delta s$. Equation 2-17 can be used to calculate the proper time interval between two events from measurements of $\Delta t$ and $\Delta x$ made in *any* inertial frame. The choice does not matter, because all observers will calculate the same result.

The proper time interval $\Delta \tau$, however, will only be equal to the measured time interval $\Delta t$ in a frame in which the two events occur at the same place, that is, in a frame in which $\Delta x = 0$; frame $S'$ in Fig. 2-4 is such a frame. In all other frames, Eq. 2-17 shows that the measured time interval will be *greater than* the proper time interval, corresponding to the slowing down of moving clocks.

If the pair of events is not timelike, that is, if $(c\ \Delta t)^2$ in Eq. 2-16 is, in fact, *less than* $(\Delta x)^2$, then no frame can be found such that the events coincide in space; $\Delta s$, calculated from Eq. 2-16, is a mathematically imaginary quantity, and no physically meaningful proper time interval can be assigned using Eq. 2-17. In such cases the term $(\Delta x)^2$ in Eq. 2-16 dominates, and the pair of events is called *spacelike*. It helps in visualizing such pairs of events to imagine them as spaced relatively far apart ($\Delta x$ large) and/or separated by a relatively short time interval ($\Delta t$ small).

We can define a *proper distance* $\Delta \sigma$ for such events from Eq. 2-16. Thus,

$$\Delta \sigma = \sqrt{-(\Delta s)^2} = \sqrt{(\Delta x)^2 - (c\ \Delta t)^2}. \tag{2-18}$$

We see that $\Delta \sigma$ is also an invariant quantity, again because of the invariance of $\Delta s$. Equation 2-18 can be used to calculate the proper distance associated with any (spacelike) pair of events from measurements of $\Delta x$ and $\Delta t$ made in *any* inertial frame. Again, the choice does not matter, because all observers will get the same answer.

The proper distance $\Delta \sigma$, however, will only be equal to the measured distance $\Delta x$ in a frame in which the measurements of the two endpoints of $\Delta x$ are made simultaneously, that is, in a reference frame in which $\Delta t = 0$. Just as proper time has no physical meaning for spacelike pairs of events, proper distance has no physical meaning for timelike pairs of events; it proves impossible to find a reference frame in which timelike events occur simultaneously.

There is no one true frame (no best frame, no fixed frame) from which to judge time, space, mass or speed. One frame is as good as the next. All of our great laws are the same in each frame. The speed of light, c, is the same in each frame, and there is no way to change this speed: no change in your speed, or in the speed of the light source, can change the speed at which light comes by you. No star, ship, man, rock, or speck of dust can reach the speed of light. This holds true for all things that can be slowed to a stop: none of these can quite match the speed of light.

If a car zooms by you at high speed, you find that its length is less (and its mass is more) than when it stands still in your frame. Also, its clock runs slow. Of course, the deal works both ways: the man in the car finds you to have more mass than when you and he are in the same frame.

If two guns (one to your right, one to your left) are fired, and you find they were fired at the same time, a man who went by at high speed may have found them not to have been fired at the same time. For him, the time span $\Delta t$ may be large, not nil.

It turns out that time and space are in some sense joined. If two guns, one here and one there, are fired the time span $\Delta t$ and the space span $\Delta x$ are linked in a strange way. Take $\sqrt{c^2(\Delta t)^2 - (\Delta x)^2}$ and note that, though the $\Delta t$ and $\Delta x$ found from some Frame A are not the same as the $\Delta t$ and $\Delta x$ found from some Frame B, yet $\sqrt{c^2(\Delta t)^2 - (\Delta x)^2}$ is the same for each frame -- in fact for all frames.

Personal correspondence from William A. Shurcliff, Cambridge, Massachusetts.

When $\Delta\tau$ is real the interval is called timelike; when $\Delta\sigma$ is real the interval is called spacelike. In the timelike region we can find a frame in which the two events occur at the same place, so that $\Delta\tau$ can be thought of as the time interval between the events in that frame. In the spacelike region we can find a frame in which the events are simultaneous, so that $\Delta\sigma$ can be thought of as the spatial interval between the events in that frame.

It is possible to find event pairs for which the two terms on the right side of Eq. 2-16 are exactly equal. Such a pair is neither timelike or spacelike but is, instead, identified as *lightlike*. The name derives from the fact that, in view of the equality just assumed, $\Delta x/\Delta t = c$. The significance of this is that if a light pulse leaves one event just as it occurs, it will arrive at the other event just as *it* occurs. As Eqs. 2-17 and 2-18 show, both the proper time $\Delta\tau$ and the proper distance $\Delta\sigma$ vanish for lightlike event pairs.

## EXAMPLE 3.

***Riding a Fast Electron.*** (*a*) An electron with a kinetic energy of 50 MeV ($= 5.0 \times 10^7$ eV), such as might be produced in a linear accelerator, can be shown (see Section 3-4) to have a speed parameter $\beta$ of 0.999949. A beam of such electrons moves along the axis of an evacuated tube that is 10 m long, measured in a reference frame $S$ fixed in the laboratory. Imagine a second frame $S'$, attached to an electron in the beam and moving with it. How long would this tube seem to be to an observer in this frame? (*b*) Repeat the calculation for a 30-GeV ($= 3.0 \times 10^{10}$-eV) electron, such as might be produced in the Stanford Linear Accelerator. Such an electron can be shown to have a speed parameter $\beta$ of 0.99999999986.

(*a*) In frame $S'$ the electron is at rest and the tube is moving and thus is contracted in length according to Eq. 2-12, or

$$\Delta x' = \Delta x \sqrt{1 - \beta^2}$$
$$= (10 \text{ m}) \sqrt{1 - (0.999949)^2}$$
$$= 0.10 \text{ m} = 10 \text{ cm}.$$

(*b*) In attempting this calculation for a 30-GeV electron, a difficulty arises in that evaluating $\beta^2$ overloads the capacity of an ordinary hand cal-

culator. For electrons in this extreme relativistic realm the quantity $(1 - \beta)$ is, in fact, both more significant and more manageable than $\beta$ itself. To take advantage of this, let us put

$$\Delta x' = \Delta x \sqrt{1 - \beta^2}$$
$$= \Delta x \sqrt{(1 + \beta)(1 - \beta)}.$$

Now "$(1 + \beta)$" can, to an extremely good approximation, be replaced by "2," and $(1 - \beta)$ is $(1 - 0.99999999986) = 0.00000000014 = 1.4 \times 10^{-10}$. Thus,

$$\Delta x' = (10 \text{ m}) \sqrt{(2)(1.4 \times 10^{-10})}$$
$$= 1.7 \times 10^{-4} \text{ m} = 0.17 \text{ mm}.$$

Relativity theory is a very practical engineering matter in the design of high-energy accelerators. If it were not properly taken into account, those machines simply would not work. Purcell [16] has referred to those engineering ventures (particle accelerators, klystrons, high-voltage television tubes, electron microscopes, global navigation systems, and so forth) in which relativistic considerations play a role, as "high-gamma engineering."

## EXAMPLE 4.

***Finding the Proper Time Interval.*** Two events are viewed by an observer fixed in an inertial reference frame $S$. They occur along the $x$ axis and are separated in space by $\Delta x$ and in time by $\Delta t$. What is the proper time interval between these events? Consider three cases:

| Event Pair | $\Delta x$ | $\Delta t$ |
|---|---|---|
| (a) | $9.0 \times 10^8$ m | 5.0 s |
| (b) | $7.5 \times 10^8$ m | 2.5 s |
| (c) | $5.0 \times 10^8$ m | 1.5 s |

(a) Let us first calculate the proper time interval by finding the proper frame for these events. Recall that the proper time interval is the time interval recorded in a frame $S'$ chosen so that both events in the pair occur at the same point when viewed from that frame. A (single) clock placed at that point reads intervals of proper time. Frame $S'$ must move, as seen by $S$, at a speed $v$ such that it covers the distance $\Delta x$ ($= 9.0 \times 10^8$ m) in a time $\Delta t$ ($= 5.0$ s); in that way observes $S'$ (and his clock) can be at both events. Thus

$$v = \frac{\Delta x}{\Delta t} = \frac{9.0 \times 10^8 \text{ m}}{5.0 \text{ s}}$$
$$= 1.8 \times 10^8 \text{ m/s} = 0.60c.$$

The time interval $\Delta t'$ ($= \Delta\tau$) read by $S'$ (on his single clock) will be related to the time interval $\Delta t$ read by $S$ (on two of his clocks, separated in space) by the time dilation formula. Thus, from Eq. 2-14$b$ we have

$$\Delta\tau = \Delta t \sqrt{1 - \beta^2}$$
$$= (5.0 \text{ s}) \sqrt{1 - (0.60)^2}$$
$$= 4.0 \text{ s}.$$

As we have seen, it is not necessary to find the (proper) frame $S'$ to calculate the proper time interval $\Delta\tau$. We can calculate it from measurements in *any* frame and will get the same answer because the proper time interval is an invariant

quantity. From Eq. 2-17, then, using measurements in frame $S$, we have

$$\Delta\tau = \sqrt{(\Delta t)^2 - \left(\frac{\Delta x}{c}\right)^2}$$
$$= \sqrt{(5.0 \text{ s})^2 - \left(\frac{9.0 \times 10^8 \text{ m}}{3.0 \times 10^8 \text{ m/s}}\right)^2}$$
$$= 4.0 \text{ s},$$

in full agreement with the preceding direct calculation. Note that event pair (a), because it possesses a physically observable proper time, constitutes what we have called a *timelike* pair.

(b) For this event pair we have

$$v = \frac{\Delta x}{\Delta t} = \frac{7.5 \times 10^8 \text{ m}}{2.5 \text{ s}}$$
$$= 3.0 \times 10^8 \text{ m/s} = c.$$

We thus see that the proper frame $S'$ would have to move at the speed of light. The event pair is *lightlike* and the proper time interval, calculated by either of the approaches used in (a), is zero.

(c) For this event pair we have

$$v = \frac{\Delta x}{\Delta t} = \frac{5.0 \times 10^8 \text{ m}}{1.5 \text{ s}}$$
$$= 3.3 \times 10^8 \text{ m/s} = 1.1c.$$

No reference frame can move so fast relative to another, so we conclude that a proper frame simply does not exist. There is no frame, that is, in which the two events would occur at the same place; they are separated in space for *all* inertial observers. Calculations of the proper time interval, carried out as in (a), would yield a mathematically imaginary result, devoid of physical meaning. We have called such event pairs *spacelike*.

---

**EXAMPLE 5.** _____

*Two Observers View Two Events.* In inertial system $S$ an event occurs on the $x$ axis at point $A$ and then, 1.0 μs ($= 1.0 \times 10^{-6}$ s) later, an event occurs at point $B$ farther out on the $x$ axis. $A$ and $B$ are 600 m apart in frame $S$. (See Fig. 2-5.) (a) Does there exist another inertial frame $S'$ in which the two events will be seen to occur

simultaneously? If so, what are the magnitude and direction of the velocity of $S'$ with respect to $S$? (b) What is the separation of events $A$ and $B$ in frame $S'$? Assume that $S$ and $S'$ are related as in Fig. 1-1. (c) What is the situation if the separation between the events in frame $S$ is 100 m, all else remaining unchanged?

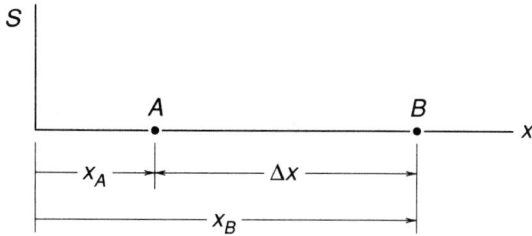

**FIGURE 2-5.**  Example 5.

(a) From the Lorentz transformation equations (Table 2-3, Eq. 4'), we have

$$\Delta t' = \gamma(\Delta t - v\,\Delta x/c^2).$$

If the events are to be simultaneous in $S'$, we must have $\Delta t' = 0$, which leads to

$$0 = \Delta t - \frac{v}{(c^2)}\,\Delta x$$

and thus to

$$v = \frac{c^2\,\Delta t}{\Delta x}$$

$$= \frac{(3.0 \times 10^8 \text{ m/s})^2(1.0 \times 10^{-6} \text{ s})}{600 \text{ m}}$$

$$= 1.5 \times 10^8 \text{ m/s} = 0.50c.$$

So, an observer in a system $S'$ moving from $A$ toward $B$ at half the speed of light would record the events as simultaneous. We have seen that a pair of events for which such a frame can be found is described as *spacelike*. You can show, us-ing the methods of Example 4, that it is not pos-sible to assign a proper time interval to this pair of events.

(b) Again, from the Lorentz transformation equations (Table 2-3; Eq. 1), we have

$$\Delta x = \gamma(\Delta x' + v\,\Delta t').$$

But the events are simultaneous in $S'$, so $\Delta t' = 0$. Thus (see Eq. 2-10), the separation of the events in $S'$ is

$$\Delta x' = \frac{\Delta x}{\gamma} = \Delta x\sqrt{1 - \beta^2}$$

$$= (600 \text{ m})\sqrt{1 - (0.50)^2} = 5.20 \text{ m}.$$

This is simply the familiar length contraction re-lationship (Eq. 2-12a), the length $AB$ ($= \Delta x$) be-ing the length of a rod at rest in frame $S$.

(c) The relative velocity $v$, worked out as in (a), proves to be

$$v = \frac{c^2\,\Delta t}{\Delta x} = \frac{(3.0 \times 10^8 \text{ m/s})^2(1.0 \times 10^{-6} \text{ s})}{100 \text{ m}}$$

$$= 9.0 \times 10^8 \text{ m/s} = 3.0c,$$

which exceeds the speed of light. Thus there is no frame in which these events would be seen as simultaneous. They occur at different times for all inertial observers. Such events are called *timelike*. You can show, using the methods of Example 4, that it is possible to calculate a proper time interval for this pair of events.

## 2-4 The Lorentz Equations—A More Physical Look

Among the most important consequences of the Lorentz transformation equations are these: (1) Lengths perpendicular to the relative motion are measured to be the same in both frames; (2) the time interval indicated on a clock is measured to be longer by an observer for whom the clock is moving than by one at rest with respect to the clock; (3) lengths parallel to the relative motion are measured to be contracted compared to the rest lengths by the observer for whom the measured bodies are moving; and (4) two clocks, which are synchronized and separated in one inertial frame, are observed to be out of synchronism from another inertial frame. Here we rederive these features one at a time by thought experiments that focus on the measuring process.

**Comparison of Lengths Perpendicular to the Relative Motion**   Imagine two frames whose relative motion $v$ is along a common $x$-$x'$ axis. In each frame an observer has a stick extending up from the origin along her vertical ($y$ and $y'$) axis, which she measures to have a (rest) length of exactly 1 m, say. As these observers approach and pass each other, we wish to determine whether or not, when the origins coincide, the top ends of the sticks coincide. We can arrange to have the sticks mark each other permanently by a thin pointer at the very top of each (for example, a razor blade or a paintbrush bristle) as they pass one another. (We displace the sticks very slightly so that they will not collide, always keeping them parallel to the vertical axis.) Notice that the situation is perfectly symmetrical. Each observer claims that her stick is a meter long, each sees the other approach with the same speed $v$, and each claims that her stick is perpendicular to the relative motion. Furthermore, the two observers must agree on the result of the measurements because they agree on the simultaneity of the measurements (the measurements occur at the instant the origins coincide). After the sticks have passed, either each observer will find her pointer marked by the other's pointer, or else one observer will find a mark below her pointer, the other observer finding no mark. That is, either the sticks are found to have the same length by both observers, or else there is an absolute result, agreed on by both observers, that the same one stick is shorter than the other. That each observer finds the other stick to be the *same* length as hers follows at once from the contradiction any other result would indicate with the relativity principle. Suppose, for example, that observer $S$ finds that the $S'$ stick has left a mark (below her pointer) on her stick. She concludes that the $S'$ stick is *shorter* than hers. This is an absolute result, for the $S'$ observer will find no mark on her stick and will conclude *also* that her stick is shorter. If, instead, the mark was left on the $S'$ stick, then *each* observer would conclude that the $S$ stick is the *shorter* one. In either case, this would give us a physical basis for preferring one frame over another, for although all the conditions are symmetrical, the results would be unsymmetrical—a result that contradicts the principle of relativity. That is, the laws of physics would not be the same in each inertial frame. We would have a property for detecting absolute motion, in this case; a shrinking stick would mean absolute motion in one direction and a stretching stick would mean absolute motion in the other direction. Hence, to conform to the relativity postulate, we conclude that the length of a body (or space interval) transverse to the relative motion is measured to be the same by all inertial observers.

**Comparison of Time-Interval Measurements**   A simple thought experiment that reveals in a direct way the quantitative relation connecting the time interval between two events as measured from two different inertial frames is the following. Imagine a passenger sitting on a train that moves with uniform velocity $v$ with respect to the ground. The experiment will consist of turning on a flashlight aimed at a mirror directly above on the ceiling and measuring the time it takes the light to travel up and be reflected back down to its starting point. The situation is illustrated in Fig. 2-6. The passenger, who has a wristwatch, sees the light ray follow a strictly vertical path (Fig. 2-6a) from $A$ to $B$ to $C$ and times the event by her clock (watch). This interval $\Delta\tau$ is a proper time interval, measured by a single clock at one place, the departure and arrival of the light ray occurring at the same place in the passen-

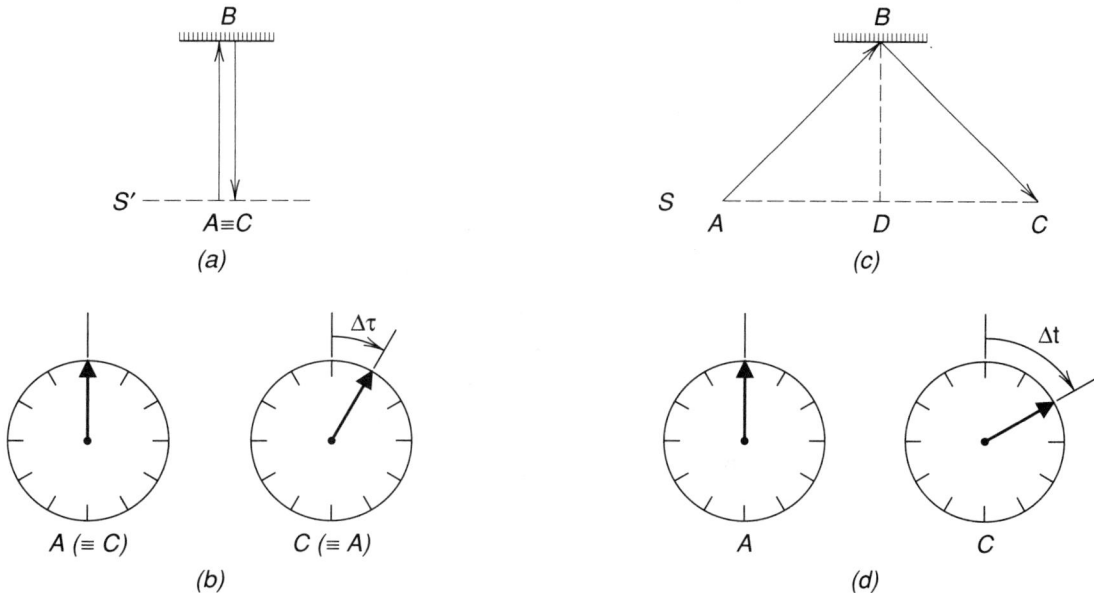

**FIGURE 2-6.** (a) The path of a light ray as seen by a passenger in the S′ frame. B is a mirror on the ceiling. A and C are the *same* point, namely, the bulb of the flashlight, in this frame. (b) The readings of the passenger's clock at the start and at the end of the experiment, showing the time interval $\Delta t'$ ($= \Delta\tau$) on the (single) S′ clock, stationary in this frame. (c) The path of a light ray as seen by a ground observer, in the S frame. A and C are *different locations* of the flashlight bulbs at the start and at the end of the experiment, as the train moves to the right with speed *v*, in this frame. (d) Readings on the *two* (synchronized) *clocks,* stationary in the S frame and located at the start (A) of the experiment and the end (C) of the experiment, showing the time interval $\Delta t$.

ger's (S′) frame. Another observer, fixed to the ground (S) frame, sees the train and passenger move to the right during this interval. He will measure the time interval $\Delta t$ from the readings on two clocks, stationary in his frame, one being at the position the experiment began (turning on of flashlight) and a second at the position the experiment ended (arrival of light to flashlight). Hence, he compares the reading of one moving clock (the passenger's watch) to the readings on two stationary clocks. For the S observer, the light ray follows the oblique path shown in Fig. 2-6c. Thus, the observer on the ground measures the light as traveling a greater distance than does the passenger (we have already seen that the transverse distance is the same for each observer). Because the speed of light is the same in both frames, the ground observer sees more time elapse between the departure and the return of the ray of light than does the passenger. He therefore concludes that the passenger's clock runs slow (see Fig. 2-6b and 2-6d). The quantitative result follows at once from the Pythagorean theorem, for

$$\Delta\tau = \frac{2BD}{c} \quad \text{and} \quad \Delta t = \frac{AB + BC}{c} = \frac{2AB}{c};$$

but

$$(BD)^2 = (AB)^2 - (AD)^2,$$

so that

$$\frac{\Delta\tau}{\Delta t} = \frac{BD}{AB} = \frac{\sqrt{(AB)^2 - (AD)^2}}{AB}$$

$$= \sqrt{1 - \left(\frac{AD}{AB}\right)^2} = \sqrt{1 - \frac{v^2}{c^2}}.$$

Here $AD$ is the horizontal distance traveled at speed $v$ during the time the light traveled with speed $c$ along the hypotenuse $AB$. This result can be written as

$$\Delta t = \frac{\Delta\tau}{\sqrt{1 - \beta^2}}$$

and is identical to Eq. 2-14b, derived earlier in a more formal way.

**Comparison of Lengths Parallel to the Relative Motion** The simplest deduction of the length contraction uses the time dilation result just obtained and shows directly that length contraction is a necessary consequence of time dilation. Imagine, for example, that two different inertial observers, one sitting on a train moving through a station with uniform velocity $v$ and the other at rest in the station, want to measure the length of the station's platform. The ground observer, for whom the platform is at rest, measures the length to be $l_0$ and claims that the passenger covered this distance in a time $l_0/v$. This time, $\Delta t$, is a nonproper time, for the events observed (passenger passes back end of platform, passenger passes front end of platform) occur at two different places in the ground ($S$) frame and are necessarily timed by two different clocks. The passenger, however, observes the platform approach and recede and finds the two events to occur at the same place in her ($S'$) frame. That is, her clock (wristwatch) is located at each event as it occurs. She measures a proper time interval $\Delta\tau$, which, as we have just seen (Eq. 2-14b), is related to $\Delta t$ by $\Delta\tau = \Delta t\sqrt{1-\beta^2}$. But $\Delta t = l_0/v$, so that $\Delta\tau = l_0\sqrt{1-\beta^2}/v$. The passenger claims that the platform moves with the same speed $v$ relative to her so that she would measure the distance from back to front of the platform as $v\,\Delta\tau$. Hence, the length of the platform to her is $l = v\,\Delta\tau = l_0\sqrt{1-\beta^2}$, which is precisely Eq. 2-12b, the length-contraction result. Thus, a body of rest length $l_0$ is measured to have a length $l_0\sqrt{1-\beta^2}$ parallel to the relative motion in a frame in which the body moves with speed $v$.

**The Phase Difference in the Synchronization of Clocks** The Lorentz transformation equation for the time variable (see Table 2-2, Eq. 4) can be written as

$$t = \gamma\left(t' + \frac{vx'}{c^2}\right) = \gamma\left(t' + \frac{\beta x'}{c}\right).$$

Here we wish to give a physical interpretation of the $\beta x'/c$ term, which we call the *phase difference*. We shall synchronize two clocks in one frame and examine what an observer in another frame concludes about the process.

Imagine that we have two clocks, $A$ and $B$, at rest in the $S'$ frame. Their separation is $L'$ in this frame. We set off a flashbulb, which is at the exact midpoint, and instruct two assistants, one at each clock, to set them to read $t' = 0$ when the light reaches them (see Fig. 2-7a). This is an agreed-upon procedure for synchronizing two separated clocks (see Section 2-1). We look at this synchronization process as seen by an observer in the $S$ frame, for whom the clocks $A$ and $B$ move to the right (see Fig. 2-7b) with speed $v$. The $S$ observer has at her disposal her own fixed array of synchronized clocks, so she can assign times of occurrence to various events.

To the $S$ observer, the separation of the two clocks will be $L'\sqrt{1-\beta^2}$. She observes the following sequence of events. The flash goes off and leaves the midpoint traveling in all directions with a speed $c$. As the wavefront expands at the rate $c$, the clocks move to the right at the rate $v$. Clock $A$ intercepts the flash first, before $B$, and the assistant at clock $A$ sets his clock at $t' = 0$ (third picture in sequence). Hence, as far as the $S$ observer is concerned, the assistant at $A$ sets his clock to zero time *before* the assistant at $B$ does, and the setting of the two primed clocks does not appear simultaneous to her. Here again we see the relativity of simultaneity; that is,

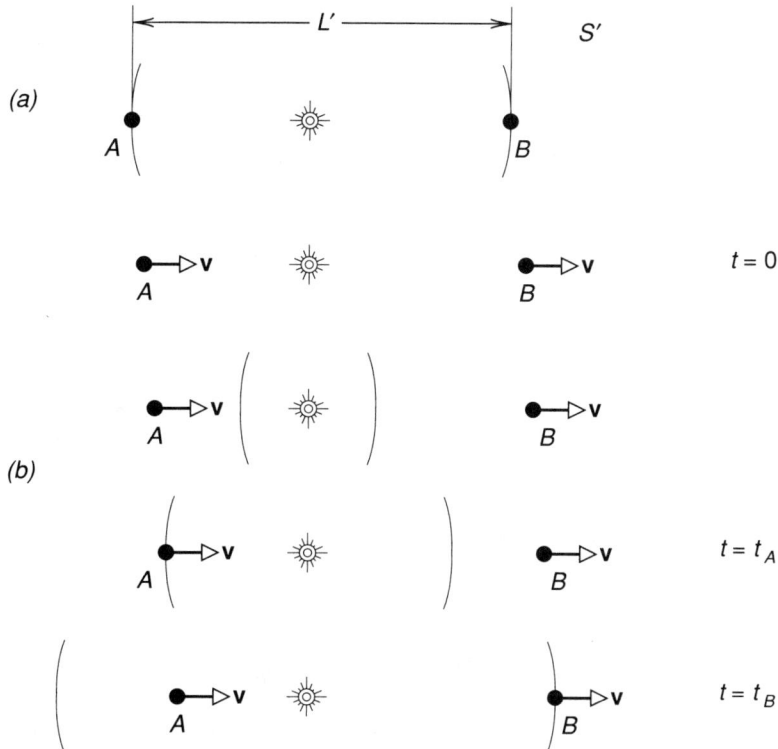

**FIGURE 2-7.** (a) A flash sent from the midpoint of clocks $A$ and $B$, at rest in the $S'$ frame a distance $L'$ apart, arrives simultaneously at $A$ and at $B$. (b) The sequence of events as seen from the $S$ frame, in which the clocks are a distance $L$ apart and move to the right with speed $v$.

the clocks at rest in the primed frame are *not* synchronized according to the unprimed observer, who uses exactly the same procedure to synchronize her own clocks.

By how much do the two $S'$ clocks differ in their readings according to the $S$ observer? Let $t = 0$ be the time $S$ sees the flash go off. Then, when the light pulse meets clock $A$, at $t = t_A$, we have

$$ct_A = (L'/2)\sqrt{1 - \beta^2} - vt_A$$

or

$$t_A = \frac{(L'/2)\sqrt{1 - \beta^2}}{c + v} = \left(\frac{L'}{2c}\right)\frac{\sqrt{1 - \beta^2}}{1 + \beta}.$$

That is, the distance the pulse travels to meet $A$ is less than their initial separation by the distance $A$ travels to the right during this time. When the light pulse later meets clock $B$ (fourth picture in sequence), at $t = t_B$, we have

$$ct_B = \frac{L'}{2}\sqrt{1 - \beta^2} + vt_B$$

or

$$t_B = \frac{(L'/2)\sqrt{1 - \beta^2}}{c - v} = \left(\frac{L'}{2c}\right)\frac{\sqrt{1 - \beta^2}}{1 - \beta}.$$

The distance the pulse travels to meet $B$ is greater than their initial separation by the distance $B$ travels to the right during this time. As measured by the clocks in $S$, therefore, the time interval between the setting of the primed clocks ($A$ and $B$) is

$$\Delta t = t_B - t_A$$

$$= \left(\frac{L'\sqrt{1 - \beta^2}}{2c}\right)\left(\frac{1}{1 - \beta} - \frac{1}{1 + \beta}\right)$$

$$= \frac{L'\beta}{c}\frac{1}{\sqrt{1 - \beta^2}}.$$

During this interval, however, $S$ observes clock $A$ to run slow by the factor $\sqrt{1 - \beta^2}$ (for "moving clocks run slow"), so to observer $S$ it will read

$$\Delta t' = \Delta t\sqrt{1 - \beta^2} = \frac{L'\beta}{c}$$

when clock $B$ is set to read $t' = 0$.

The result is that the $S$ observer finds the $S'$ clocks to be out of synchronization, with clock $A$ reading *ahead* in time by an amount $L'\beta/c$ ($= x'\beta/c$). The greater the separation $L'$ of the clocks in the primed frame, the further behind in time is the reading of the $B$ clock as observed at a given instant from the unprimed frame. This is in exact agreement with the Lorentz transformation equation for the time.

Hence, all the features of the Lorentz transformation equations, which we derived in a formal way directly from the postulates of relativity in Section 2-2, can be derived more physically from the measurement processes, which were, of course, chosen originally to be consistent with those postulates.

# EXAMPLE 6.

*Simultaneity at Low Speeds.* Why is the fact that simultaneity is not an absolute concept an unexpected result to the classical mind? It is because the speed of light has such a large value compared to ordinary speeds.

Consider these two cases, which are symmetrical in terms of an interchange of the space and time coordinates. *Case 1:* $S'$ observes that two events occur at the same place but are separated in time; $S$ will then declare that the two events occur in different places. *Case 2:* $S'$ observes that two events occur at the same time but are separated in space; $S$ will then declare that the two events occur at different times.

Case 1 is readily acceptable on the basis of daily experience. If a person ($S'$) on a moving train winks and then—ten minutes later—winks again, these events occur at the same place on *his* reference frame (the train). A ground observer ($S$), however, would assert that these same events occur at different places in *his* reference system (the ground). Case 2, although true, cannot be easily supported on the basis of daily experience. Suppose that $S'$, seated at the center of a moving railroad car, observes that two children, one at each end of the car, wink simultaneously. The ground observer $S$, watching the railroad car go by, would assert (if he could make precise enough measurements) that the child in the back of the car winked a little before the child in the front of the car did. The fact that the speed of light is so high compared to the speeds of familiar large objects makes Case 2 less intuitively reasonable than Case 1, as we now show.

(*a*) In Case 1, assume that the time separation in $S'$ is 10 min; what is the distance separation observed by $S$? (*b*) In Case 2, assume that the distance separation in $S'$ is 25 m; what is the time separation observed by $S$? Take $v = 20.0$ m/s,

which corresponds to 45 mi/h or $\beta = v/c = 6.7 \times 10^{-8}$.

(*a*) From Table 2-3, Eq. 1 we have

$$x_2 - x_1 = \frac{x'_2 - x'_1}{\sqrt{1 - \beta^2}} + \frac{v(t'_2 - t'_1)}{\sqrt{1 - \beta^2}}.$$

We are given that $x'_2 = x'_1$ and $t'_2 - t'_1 = 10$ min, so

$$x_2 - x_1 = \frac{(20.0 \text{ m/s})(600 \text{ s})}{\sqrt{1 - (6.7 \times 10^{-8})^2}}$$
$$= 12000 \text{ m} = 12 \text{ km}.$$

This result is readily accepted. Because the denominator above is unity for all practical purposes, the result is even numerically what we would expect from the Galilean equations.

(*b*) From Table 2-3, Eq. 4 we have

$$t_2 - t_1 = \frac{t'_2 - t'_1}{\sqrt{1 - \beta^2}} + \frac{(v/c^2)(x'_2 - x'_1)}{\sqrt{1 - \beta^2}}.$$

We are given that $t'_2 = t'_1$ and that $x'_2 - x'_1 = 25$ m, so

$$t_2 - t_1 = \frac{[(20 \text{ m/s})/(3.0 \times 10^8 \text{ m/s})^2](25 \text{ m})}{\sqrt{1 - (6.7 \times 10^{-8})^2}}$$
$$= 5.6 \times 10^{-15} \text{ s}.$$

The result is *not* zero, a value that would have been expected by classical physics, but the time interval is so short that it would be very hard to show experimentally that it really was not zero.

If we compare the expressions for $x_2 - x_1$ and for $t_2 - t_1$ above, we see that, whereas $v$ appears as a factor in the second term of the former, $v/c^2$ appears in the latter. Thus the relatively high value of $c$ puts Case 1 within the bounds of familiar experience but puts Case 2 out of these bounds.

---

In the following example we consider the realm wherein relativistic effects are easily observable.

## EXAMPLE 7

*The Decay of Moving Pions.* Among the particles of high-energy physics are charged pions, particles of mass between that of the electron and the proton and of positive or negative electronic charge. They can be produced by bombarding a suitable target in an accelerator with high-energy protons, the pions leaving the target with speeds close to that of light. It is found that the pions are radioactive and, when they are brought to rest, their half-life is measured to be $1.8 \times 10^{-8}$ s. That is, half of the number present at any time have decayed $1.8 \times 10^{-8}$ s later. A collimated pion beam, leaving the accelerator target at a speed of $0.99c$, is found to drop to half its original intensity 38 m from the target.

(*a*) Are these results consistent?

If we take the half-life to be $1.8$ s $\times 10^{-8}$ s and the speed to be $2.97 \times 10^8$ m/s ($= 0.99c$), the distance traveled over which half the pions in the beam should decay is

$$d = vt = 2.97 \times 10^8 \text{ m/s} \times 1.8 \times 10^{-8} \text{ s}$$
$$= 5.3 \text{ m.}$$

This appears to contradict the direct measurement of 38 m.

(*b*) Show how the time dilation accounts for the measurements.

If the relativistic effects did not exist, then the half-life would be measured to be the same for pions at rest and pions in motion (an assumption we made in part *a*). In relativity, however, the nonproper and proper half-lives are related by Eq. 2-14*b*, or

$$\Delta t = \frac{\Delta \tau}{\sqrt{1 - \beta^2}}.$$

The proper time in this case is $1.8 \times 10^{-8}$ s, the time interval measured by a clock attached to the pion, that is, at one place in the rest frame of the pion. In the laboratory frame, however, the pions are moving at high speeds and the time interval there (a nonproper one) will be measured to be larger (moving clocks appear to run slow). The non-proper half-life, measured by two different clocks in the laboratory frame, would then be

$$\Delta t = \frac{1.8 \times 10^{-8} \text{ s}}{\sqrt{1 - (0.99)^2}} = 1.28 \times 10^{-7} \text{ s.}$$

This is the half-life appropriate to the laboratory reference frame. Pions that live this long, traveling at a speed $0.99c$, would cover a distance

$$d = 0.99c \times \Delta t$$
$$= 2.97 \times 10^{-8} \text{ m/s} \times 1.28 \times 10^{-7} \text{ s}$$
$$= 38 \text{ m,}$$

exactly as measured in the laboratory.

(*c*) Show how the length contraction accounts for the measurements.

In part *a* we used a length measurement (38 m) appropriate to the laboratory frame and a time measurement ($1.8 \times 10^{-8}$ s) appropriate to the pion frame and incorrectly combined them. In part *b* we used the length (38 m) and time ($1.28 \times 10^{-7}$ s) measurements appropriate to the laboratory frame. Here we use length and time measurements appropriate to the pion frame.

We already know the half-life in the pion frame, that is, the proper time $1.8 \times 10^{-8}$ s. What is the distance covered by the pion beam during which its intensity falls to half its original value? If we were sitting on the pion, the laboratory distance of 38 m would appear much shorter to us because the laboratory moves at a speed $0.99c$ relative to us (the pion). In fact, we would measure the distance

$$d' = d \sqrt{1 - \beta^2} = 38 \sqrt{1 - (0.99)^2} \text{ m.}$$

The time elapsed in covering this distance is $d'/0.99c$ or

$$\Delta \tau = \frac{38 \text{ m} \sqrt{1 - (0.99)^2}}{0.99c} = 1.8 \times 10^{-8} \text{ s,}$$

exactly the measured half-life in the pion frame.

Thus, depending on which frame we choose to make measurements in, this example illustrates the physical reality of either the time-dilation or the length-contraction predictions of relativity. Each pion carries its own clock, which determines the proper time $\tau$ of decay, but the decay time observed by a laboratory observer is much greater. Or, expressed equivalently, the moving pion sees the laboatory distances contracted and in its proper decay time can cover laboratory distances greater than those measured in its own frame.

Notice that in this region of $v \cong c$ the relativistic effects are large. There can be no doubt whether, in our example, the distance is 38 m or 5.3 m. If the proper time were applicable to the laboratory frame, the time $(1.28 \times 10^{-7}$ s$)$ to travel 38 m would correspond to more than seven half-lives (that is, $1.28 \times 10^{-7}$ s/$1.8 \times 10^{-8}$ s $\cong 7$). Instead of the beam being reduced to half its original intensity, it would be reduced to $(1/2)^7$ or 1/128 its original intensity in travelling 38 m. Such differences are very easily detectable. See Problems 23-27 and also [2] for other examples in which relativistic considerations are a central part of the problem at hand.

---

## EXAMPLE 8.

*Two Spaceships Pass Each Other.* Two spaceships, each of proper length 100 m, pass near one another heading in opposite directions; see Fig. 2-8. An astronaut at the front of one ship ($S$) measures a time interval of $2.50 \times 10^{-6}$ s for the second ship ($S'$) to pass her. (*a*) What is the relative speed $v$ of the two ships? (*b*) What time interval is measured on ship $S$ for the front of ship $S'$ to pass from the front to the back of $S$? In the figure the front and back ends of spaceship $S$ are labeled $A$ and $B$, respectively, those of spaceship $S'$ being labeled $A'$ and $B'$. Let $AA'$ mean the coincidence of points $A$ and $A'$, $AB'$ the coincidence of $A$ and $B'$, and so forth.

(*a*) The time interval between the occurrences of events $AA'$ and $AB'$, measured by a single clock at A, is a proper time interval and is given as $2.50 \times 10^{-6}$ s $(= \Delta\tau)$. Each ship has a proper length of 100 m $(= L_0)$ so the space interval between $A'$ and $B'$ that the astronaut at $A$ measures is the contracted length $L_0\sqrt{1-\beta^2}$, appropriate, as Eq. 2-12b shows, for an object of rest length $L_0$ moving at a speed $v$. Therefore,

$$v(= \beta c) = \frac{L_0\sqrt{1-\beta^2}}{\Delta\tau}$$

or

$$\left(\frac{\beta\Delta\tau c}{L_0}\right)^2 = 1 - \beta^2.$$

This can be written as

$$\frac{1}{\beta^2} = \left(\frac{\Delta\tau c}{L_0}\right)^2 + 1$$

$$= \left[\frac{(2.50 \times 10^{-6} \text{ s}) (3.00 \times 10^8 \text{ m/s})}{100 \text{ m}}\right]^2 + 1$$

$$= 57.25,$$

which yields

$$\beta = \frac{1}{\sqrt{57.25}} = 0.132$$

and

$$v = \beta c = (0.132)(3.00 \times 10^8 \text{ m/s})$$

$$= 3.96 \times 10^7 \text{ m/s}.$$

(*b*) We want to find the time interval between events $AA'$ and $BA'$ measured by two clocks in spaceship $S$, one at $A$ and one at $B$. This is a nonproper time interval $\Delta t$, read off as the difference in arrival times of $A'$ at the clocks at $A$ and $B$. Since the separation of these clocks is $L_0$ in spaceship $S$ and $A'$ moves at speed $v$ relative to this ship, we have

$$\Delta t = \frac{L_0}{v} = \frac{100 \text{ m}}{3.96 \times 10^7 \text{ m/s}}$$

$$= 2.53 \times 10^{-6} \text{ s}.$$

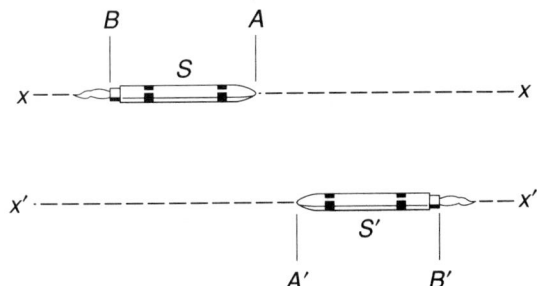

FIGURE 2-8.   Example 8.

## 2-5 The Observer in Relativity

There are many shorthand expressions in relativity that can easily be misunderstood by the uninitiated. Thus the phrase "moving clocks run slow" means that a clock moving at a constant velocity relative to an inertial frame containing synchronized clocks will be found to run slow *when timed by those clocks*. We compare *one moving clock* with *two synchronized stationary clocks*. Those who assume that the phrase means anything else often encounter difficulties.

Similarly, we often refer to "an observer." The meaning of this term also is quite definite, but it can be misinterpreted. *An observer is really an infinite set of recording clocks distributed throughout space, at rest and synchronized with respect to one another.* The space-time coordinates of an event $(x, y, z, t)$ are recorded by the clock at the location $(x, y, z)$ of the event at the time $(t)$ it occurs. Measurements thus recorded throughout space-time (we might call them local measurements) are then available to be picked up and analyzed by an experimenter. Thus, the observer can also be thought of as the experimenter who collects the measurements made in this way. Each inertial frame is imagined to have such a set of recording clocks, or such an observer. The relations between the space-time coordinates of a physical event measured by one observer $(S)$ and the space-time coordinates of the *same* physical event measured by another observer $(S')$ are the equations of transformation.

A misconception of the term "observer" arises from confusing "measuring" with "seeing." For example, it had been commonly assumed for some time that the relativistic length contraction would cause rapidly moving objects to appear to the eye to be shortened in the direction of motion. The location of all points of the object measured at the same time would give the "true" picture according to our use of the term "observer" in relativity. But, in the words of V. F. Weisskopf [Ref. 3]:

> When we see or photograph an object, we record light quanta emitted by the object when they arrive simultaneously at the retina or at the photographic film. This implies that these light quanta have *not* been emitted simultaneously by all points of the object. The points further away from the observer have emitted their part of the picture earlier than the closer points. Hence, if the object is in motion, the eye or the photograph gets a distorted picture of the object, since the object has been at different locations when different parts of it have emitted the light seen in the picture.

To make a comparison with the relativistic predictions, therefore, we must first allow for the time of flight of the light quanta from the different parts of the object. Without this correction, we see a distortion due to *both* the optical *and* the relativistic effects. Circumstances sometimes exist in which the object appears to have suffered no contraction at all. Under other special circumstances the Lorentz contraction can be seen unambiguously (see Refs. 4 and 5). But the term "observer" does *not* mean "viewer" in relativity, and we shall continue to use it only in the sense of "measurer" described above.

## 2-6 The Relativistic Addition of Velocities

In classical physics, if we have a train moving with a velocity **v** with respect to the ground and a passenger on the train moves with a velocity **u'** with respect to the train, then the passenger's velocity relative to the ground **u** is just the vector sum of the two velocities (see Eq. 1-5); that is,

$$\mathbf{u} = \mathbf{u'} + \mathbf{v}. \tag{2-19}$$

This is simply the classical, or Galilean, velocity addition theorem (See *Physics, Part I*, Sec. 4-6). How do the velocities add in special relativity theory?

Consider, for the moment, the special case wherein all velocities are along the common $x$-$x'$ direction of two inertial frames $S$ and $S'$. Let $S$ be the ground frame and $S'$ the frame of the train, whose speed relative to the ground is $v$ (see Fig. 2-9). A passenger is walking along the aisle toward the front of the train with a speed $u'$ relative to the train. His position on the train as time goes on can be described by $x' = u't'$. What is the speed of the passenger observed from the ground? Using the Lorentz transformation equations (Table 2-2, Eqs. 1' and 4'), we have

$$x' = \gamma(x - vt) = u't' \quad \text{and} \quad t' = \gamma\left(t - \frac{vx}{c^2}\right).$$

Combining these yields

$$x - vt = u'\left(t - \frac{vx}{c^2}\right),$$

which can be written as

$$x = \frac{(u' + v)}{(1 + u'v/c^2)}\, t. \tag{2-20}$$

If we call the passenger's speed relative to ground $u$, then his ground location as time goes on is given by $x = ut$. Comparing this to Eq. 2-20, we obtain

$$u = \frac{u' + v}{1 + u'v/c^2}. \tag{2-21}$$

This is the *relativistic*, or Einstein *velocity addition theorem*.

Note that if $u' = 0$, Eq. 2-21 gives $u = v$, an expected result if the passenger stops walking. If $v = 0$, we find that $u = u'$, also an expected result if the train stops. If $u'$ and $v$ are very small compared to $c$, Eq. 2-21 reduces to the classical result, Eq. 2-19, $u = u' + v$, for then the second term in the denominator of Eq. 2-21 is negligible compared to one. On the other hand, if $u' = c$, it always follows that $u = c$ no matter what is the value of $v$. Of course, $u' = c$ means that our "passenger" is a light pulse, and we know that an assumption used to derive the transformation

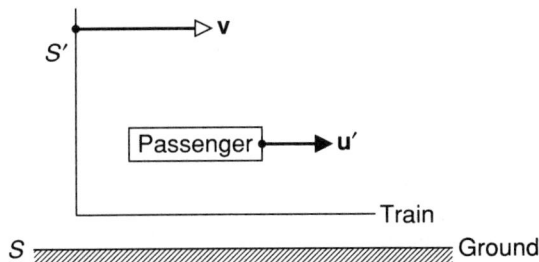

**FIGURE 2-9.** A schematic view of the system used in deriving the equations for the relativistic addition of velocities.

formulas was exactly this result, that is, that all observers measure the same speed $c$ for light. Formally, we get, with $u' = c$,

$$u = \frac{c + v}{1 + cv/c^2} = \frac{c + v}{c(c + v)}c^2 = c.$$

Hence, any velocity (less than $c$) relativistically added to $c$ gives a resultant $c$. In this sense, $c$ plays the same role in relativity that an infinite velocity plays in the classical case.

The Einstein velocity addition theorem can be used to explain the observed result of the experiments designed to test the various emission theories of Chapter 1. The basic result of these experiments is that the velocity of light is independent of the velocity of the source (see Section 1-8). We have seen that this is a basic postulate of relativity, so we are not surprised that relativity yields agreement with these experiments. If, however, we merely looked at the formulas of relativity, unaware of their physical origin, we could obtain this specific result from the velocity addition theorem directly. Let the source be the $S'$ frame. In that frame the pulse (or wave) of light has a speed $c$ in vacuum according to the emission theories. Then, the pulse (or wave) speed measured by the $S$ observer, for whom the source moves, is given by Eq. 2-21, and is also $c$. That is, $u = c$ when $u' = c$, as shown above.

It follows also from Eq. 2-21 that the addition of two velocities, each smaller than $c$, cannot exceed the velocity of light.

N. David Mermin has given a simple and convincing proof of Eq. 2-21 whose only nonclassical feature is the postulate of the constancy of the speed of light. The proof follows directly from this postulate and does not invoke either the Lorentz transformation equations, or the time dilation or the length contraction results [15].

## EXAMPLE 9.

***The Relative Speed of Two Fast Electrons.*** In Example 3 of Chapter 1, we found that when two electrons leave a radioactive sample in opposite directions, each having a speed $0.67c$ with respect to the sample, the speed of one electron relative to the other is $1.34c$ according to classical physics. What is the relativistic result?

We may regard one electron as the $S$ frame, the sample as the $S'$ frame, and the other electron as the object whose speed in the $S$ frame we seek (see Fig. 1-3). Then

$$u' = 0.67c, \quad v = 0.67c$$

$$\text{and } u = \frac{u' + v}{1 + u'v/c^2} = \frac{(0.67 + 0.67)c}{1 + (0.67)^2}$$

$$= \frac{1.34}{1.45}c = 0.92c.$$

The speed of one electron relative to the other is less than $c$.

Does the relativistic velocity addition theorem alter the numerical result of Example 2 of Chapter 1? Explain.

## EXAMPLE 10.

***Relativity Explains the Fresnel Drag Coefficient.*** Show that the Einstein velocity addition theorem leads to the observed Fresnel drag coefficient of Eq. 1-12.

In this case, $v_w$ is the velocity of water with respect to the apparatus and $c/n$ is the velocity of

light relative to the water. That is, in Eq. 2-21, we have

$$u' = \frac{c}{n} \quad \text{and} \quad v = v_w.$$

Then, the velocity of light relative to the apparatus is

$$u = \frac{c/n + v_w}{1 + v_w/nc}.$$

For $v_w/nc \ll 1$ (in the experiment its value was $1.8 \times 10^{-8}$), we can neglect terms of second-order in this quantity so that, using the binomial expansion, we have

$$u = \left(\frac{c}{n} + v_w\right)\left(1 + \frac{v_w}{nc}\right)^{-1}$$

$$= \left(\frac{c}{n} + v_w\right)\left(1 - \frac{v_w}{nc} + \cdots\right)$$

$$\cong \frac{c}{n} + v_w\left(1 - \frac{1}{n^2}\right).$$

This is exactly Eq. 1-12, the observed first-order effect. Notice that there is no need to assume any "drag" mechanism, or to invent theories on the interaction between matter and the "ether." The result is an inevitable consequence of the velocity addition theorem and illustrates the powerful simplicity of relativity.

It is interesting and instructive to note that there *are* speeds in excess of $c$. Although matter or energy (that is, signals) cannot have speeds greater than $c$, certain kinematical processes *can* have superlight speeds [6]. For example, the succession of points of intersection of the blades of a giant scissors, as the scissors is rapidly closed, may be generated at a speed greater than $c$. Here geometric points are involved, the motion being an illusion, whereas the material objects involved (atoms in the scissors blades, for example) always move at speeds less than $c$. Other similar examples are the succession of points on a fluorescent screen as an electron beam sweeps across the screen, or the light of a searchlight beam sweeping across the cloud cover in the sky. The electrons, or the light photons, which carry the energy, move at speeds not exceeding $c$. There is no contradiction with relativity theory in any of these situations.

It has been proposed [7] that there can exist, or be created, particles with speeds *always* greater than $c$. These hypothetical particles were named *tachyons*, from the Greek word for swift. This suggestion had the appeal of symmetry; that is, the existence of tachyons would allow us to classify particles by speed: normal particles that travel with $v < c$ always; photons and massless particles, for which $v = c$ always; and particles (tachyons) with $v > c$ always. Objections that infinite energy would be needed to create such particles, and certain causal paradoxes, can be resolved so that there are no compelling arguments against the existence of tachyons. Experimental evidence to date, however, suggests that their existence is unlikely.

Thus far, we have considered only the transformation of velocities parallel to the direction of relative motion of the two frames of reference (the $x$-$x'$ direction). To signify this, we should put $x$ subscripts on $u$ and $u'$ in Eq. 2-21, obtaining

$$u_x = \frac{u'_x + v}{1 + u'_x(v/c^2)}. \tag{2-22a}$$

For velocity components perpendicular to the direction of relative motion the result is more involved. Imagine that an object is observed to be at positions $y_1$ and $y_2$ in frame $S$ at times $t_1$ and $t_2$, respectively. Its $y$ component of velocity in $S$ is then $u_y = (y_2 - y_1)/(t_2 - t_1)$ or $\Delta y/\Delta t$. To find its $y$ component of velocity in frame $S'$, we start from the Lorentz transformation equations (Table 2-3, Eqs. 2 and 4), writing

$$\Delta y = \Delta y'$$

and

$$\Delta t = \gamma \left( \Delta t' + \frac{v\,\Delta x'}{c^2} \right)$$

so that

$$u_y = \frac{\Delta y}{\Delta t} = \frac{\Delta y'}{\gamma(\Delta t' + v\,\Delta x'/c^2)}.$$

Substituting for $\gamma$ and rearranging leads to

$$u_y = \frac{(\Delta y'/\Delta t')\sqrt{1 - v^2/c^2}}{1 + (v/c^2)(\Delta x'/\Delta t')}.$$

But

$$\frac{\Delta y'}{\Delta t'} = u'_y \quad \text{and} \quad \frac{\Delta x'}{\Delta t'} = u'_x,$$

so

$$u_y = \frac{u'_y \sqrt{1 - v^2/c^2}}{1 + vu'_x/c^2}, \tag{2-22b}$$

which is the relationship we seek.

In just the same way we find, for $u_z$,

$$u_z = \frac{u'_z \sqrt{1 - v^2/c^2}}{1 + vu'_x/c^2}. \tag{2-22c}$$

In Table 2-4 we summarize the relativistic velocity transformation equations. The inverse relations were found by merely changing $v$ to $-v$ and interchanging the primed and unprimed quantities. We shall have occasion to use these results, and to interpret them further, in later sections. For the moment, however, let us note certain aspects of the transverse velocity transformations. The perpendicular, or transverse, components (that is, $u_y$ and $u_z$) of the velocity of an object as seen in the $S$ frame are related both to the transverse components (that is, $u'_y$ and $u'_z$) and to the parallel component (that is, $u'_x$) of the velocity of the object in the $S'$ frame. The result is simple because neither observer is a proper one. If we choose a frame in which $u'_x = 0$, however, then the transverse results become $u_z = u'_z\sqrt{1 - v^2/c^2}$ and $u_y = u'_y\sqrt{1 - v^2/c^2}$. But no length contraction is involved for transverse space intervals,

**TABLE 2-4**
**The Relativistic Velocity Transformation Equations**

| | |
|---|---|
| $u'_x = \dfrac{u_x - v}{1 - u_x v/c^2}$ | $u_x = \dfrac{u'_x + v}{1 + u'_x v/c^2}$ |
| $u'_y = \dfrac{u_y \sqrt{1 - v^2/c^2}}{1 - u_x v/c^2}$ | $u_y = \dfrac{u'_y \sqrt{1 - v^2/c^2}}{1 + u'_x v/c^2}$ |
| $u'_z = \dfrac{u_z \sqrt{1 - v^2/c^2}}{1 - u_x v/c^2}$ | $u_z = \dfrac{u'_z \sqrt{1 - v^2/c^2}}{1 + u'_x v/c^2}$ |

so what is the origin of the $\sqrt{1 - v^2/c^2}$ factor? We need only point out that velocity, being a ratio of length interval to time interval, involves the time coordinate too, so time dilation is involved. Indeed, this special case of the transverse velocity transformation is a direct time-dilation effect.

## 2-7 Aberration and Doppler Effect in Relativity

Up to now we have shown how relativity can account for the experimental results of various light-propagation experiments listed in Table 1-2 (for example, the Fresnel drag coefficient and the Michelson-Morley result) and at the same time how it predicts new results also confirmed by experiment (time dilation in the decay of pions or other mesons, also in Table 1-2). Here we deduce the aberration result described in Section 1-7. In doing this, we shall also come upon another new result predicted by relativity and confirmed by experiment, namely, a transverse Doppler effect.

Consider a train of plane monochromatic light waves of unit amplitude emitted from a source at the origin of the $S'$ frame, as shown in Fig. 2-10. The rays, or wave normals, are chosen to be in (or parallel to) the $x'$-$y'$ plane, making an angle $\theta'$ with the $x'$ axis. An expression describing the propagation would be of the form

$$\cos 2\pi \left( \frac{x' \cos \theta' + y' \sin \theta'}{\lambda'} - v't' \right), \qquad (2\text{-}23)$$

for this is a single periodic function, amplitude unity, representing a wave moving with velocity $\lambda'v'$ $(= c)$ in the $\theta'$ direction. Notice, for example, that for $\theta' = 0$ it reduces to $\cos 2\pi(x'/\lambda' - v't')$, and for $\theta' = 90°$ it reduces to $\cos 2\pi$ $(y'/\lambda' - v't')$, well-known expressions for propagation along the positive $x'$ and positive $y'$ directions, respectively, of waves of frequency $v'$ and wavelength $\lambda'$.

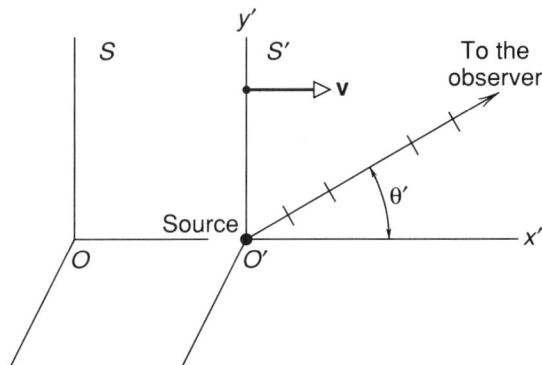

**FIGURE 2-10.** A ray, or wave normal, of plane monochromatic light waves is emitted from the origin of the $S'$ frame. The bars signify wavefronts separated by one wavelength from adjacent wavefronts. The direction of propagation makes an angle $\theta'$ with the $x'$ axis, the rays being parallel to the $x'$-$y'$ plane. The source is at rest in the $S'$ frame, and the observer is at rest in the $S$ frame.

Analysis of the quantities in the parentheses of these last two equations shows [see *Physics*, Part I, Section 19-3] that the wave speed is indeed $\lambda' v'$ which, for electromagnetic waves in free space, is equal to $c$.

In the $S$ frame these wavefronts will still be planes, for the Lorentz transformation is linear so a plane transforms into a plane. Hence, in the unprimed, or $S$, frame, the expression describing the propagation will have the same form:

$$\cos 2\pi\left(\frac{x \cos \theta + y \sin \theta}{\lambda} - vt\right), \tag{2-24}$$

Here, $\lambda$ and $v$ are the wavelength and frequency, respectively, measured by an observer fixed in the $S$ frame, and $\theta$ is the angle a ray makes with the $x$ axis. We know, if expressions 2-23 and 2-24 are to represent electromagnetic waves, that $\lambda v = c$, just as $\lambda' v' = c$, for $c$ is the velocity of electromagnetic waves, the same for each observer.

Now let us apply the Lorentz transformation equations directly to expression 2-23, putting

$$x' = \frac{x - vt}{\sqrt{1 - \beta^2}}, \quad y' = y, \quad \text{and} \quad t' = \frac{t - (v/c^2)x}{\sqrt{1 - \beta^2}}.$$

We obtain

$$\cos 2\pi\left[\frac{1}{\lambda'}\frac{(x - vt)}{\sqrt{1 - \beta^2}} \cos \theta' + \frac{y \sin \theta'}{\lambda'} - v'\frac{[t - (v/c^2)x]}{\sqrt{1 - \beta^2}}\right]$$

or, on rearranging terms and using $\lambda' v' = c$,

$$\cos 2\pi\left[\frac{\cos \theta' + \beta}{\lambda'\sqrt{1 - \beta^2}}x + \frac{\sin \theta'}{\lambda'}y - \frac{(1 + \beta \cos \theta')v'}{\sqrt{1 - \beta^2}}t\right].$$

As expected, this has the form of a plane wave in the $S$ frame and must be identical to expression 2-24, which represents the same thing. Hence, the coefficient of $x$, $y$, and $t$ in each expression must be equated, giving us

$$\frac{\cos \theta}{\lambda} = \frac{\cos \theta' + \beta}{\lambda'\sqrt{1 - \beta^2}}, \tag{2-25}$$

$$\frac{\sin \theta}{\lambda} = \frac{\sin \theta'}{\lambda'}, \tag{2-26}$$

$$v = \frac{v'(1 + \beta \cos \theta')}{\sqrt{1 - \beta^2}}. \tag{2-27}$$

We also have the relation

$$\lambda v = \lambda' v' = c, \tag{2-28}$$

a condition we knew in advance.

In the procedure we have adopted here, we start with a light wave in $S'$ for which we know $\lambda'$, $v'$, and $\theta'$ and we wish to find what the corresponding quantities $\lambda$, $v$, and $\theta$ are in the $S$ frame. That is, we have three unknowns, but we have four equations (Eqs. 2-25 to 2-28) from which to determine the unknowns. The unknowns have been overdetermined, which means simply that the equations are not all independent. If we eliminate one equation, for instance, by dividing one by another (that is, we combine two equations), we shall obtain three independent relations. It is simplest to divide Eq. 2-26 by Eq. 2-25; this gives us

$$\tan \theta = \frac{\sin \theta' \sqrt{1 - \beta^2}}{\cos \theta' + \beta}, \qquad (2\text{-}29a)$$

which is *the relativistic equation for the aberration of light*. It relates the directions of propagation, $\theta$ and $\theta'$, as seen from two different inertial frames. The inverse transformation can be written at once as

$$\tan \theta' = \frac{\sin \theta \sqrt{1 - \beta^2}}{\cos \theta - \beta}, \qquad (2\text{-}29b)$$

wherein $\beta$ of Eq. 2-29a becomes $-\beta$ and we interchange primed and unprimed quantities. Experiments in high-energy physics involving photon emission from high-velocity particles confirm the relativistic formula exactly.

## EXAMPLE 11.

*Relativity and Stellar Aberration.* Show that the classical expression for the aberration effect for an overhead star (Eq. 1-11) is an excellent first approximation to the correct relativistic expression (Eqs. 2-29).

In the $S$ frame (attached to the sun) let the one direction of propagation of light from the star be along the negative $y$ direction. Hence $\theta = 270°$. In $S'$ (attached to the earth), the propagation direction is $\theta'$, given by Eq. 2-29b with $\theta = 270°$. That is

$$\tan \theta' = \frac{\sin 270° \sqrt{1 - \beta^2}}{\cos 270° - \beta} = \frac{\sqrt{1 - \beta^2}}{\beta}.$$

When $v$ is much less than $c$, we have $\beta \ll 1$. Thus $\beta^2$ will be negligible compared to one. Neglecting terms of the second order, we can write

$$\tan \theta' = \frac{\sqrt{1 - \beta^2}}{\beta} \cong \frac{1}{\beta}$$

as a first approximation to the exact relativistic result.

Let us now replace $\theta'$ in this relationship by $\theta - \alpha$, where $\theta = 270°$ and $\alpha$ is the (very small) aberration angle displayed in Fig. 1-9 and Fig. 2-11. By trigonometry (see Problem 77), we can

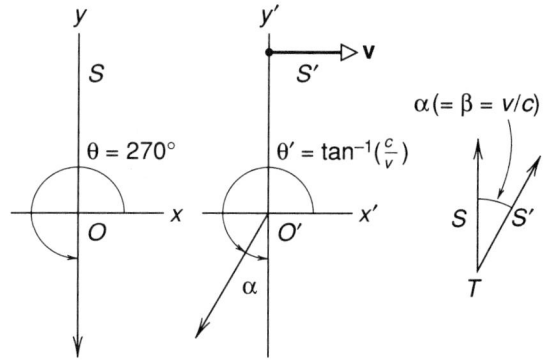

**FIGURE 2-11.** (a) In $S$, the direction of propagation from the source is the negative $y$ direction ($\theta = 270°$). In $S'$, the same ray makes an angle $\theta'$ with the $x'$ axis. (b) The line of sight of the telescope in $S$ is vertical and in $S'$ is inclined forward by an angle $\alpha(= \beta = v/c)$ in order to see the source.

easily show that the relation $\tan \theta' = (1/\beta)$ is exactly equivalent to the relation $\tan \alpha = \beta$, which (see Eq. 1-11) is the prediction of classical theory. Thus the exact classical formula does indeed appear as a first-order approximation to the relativistic formula, a not unexpected result.

The third of our four equations above (Eqs. 2-25 to 2-28) gives us directly the one remaining phenomenon we promised to discuss; that is, *the relativistic equation for the Doppler effect,*

$$\nu = \frac{\nu'(1 + \beta \cos \theta')}{\sqrt{1 - \beta^2}}, \tag{2-27a}$$

which we can also write inversely as

$$\nu' = \frac{\nu(1 - \beta \cos \theta)}{\sqrt{1 - \beta^2}}.$$

We shall find it useful to recast this as

$$\nu = \nu' \frac{\sqrt{1 - \beta^2}}{1 - \beta \cos \theta}. \tag{2-27b}$$

Let us first consider the special case of $\theta = 0$, which corresponds, as Fig. 2-11 shows, to a wave propagating in the positive direction of the $x$-$x'$ axis and to the source and the observer *approaching* each other. For $\theta = 0$, then, Eq. 2-27b becomes

$$\nu = \nu_0 \frac{\sqrt{1 - \beta^2}}{1 - \beta} = \nu_0 \sqrt{\frac{1 + \beta}{1 - \beta}} \quad \text{(approaching)}. \tag{2-30a}$$

Note that we have made a small change in notation, replacing $\nu'$, which is the frequency (a *proper* frequency) measured in a frame in which the source is at rest, by $\nu_0$.

If the source and the observer are *separating* from each other, we can find the Doppler-shifted frequency by putting $\theta = 180°$ in Eq. 2-27b or, equivalently, by changing the sign of $\beta$ in Eq. 2-30a. Either choice leads to

$$\nu = \nu_0 \frac{\sqrt{1 - \beta^2}}{1 + \beta} = \nu_0 \sqrt{\frac{1 - \beta}{1 + \beta}} \quad \text{(separating)}. \tag{2-30b}$$

For moving terrestrial light sources of macroscopic dimensions it will always be true that $v \ll c$ or, equivalently, that $\beta \ll 1$. Under these circumstances it is instructive to expand Eqs. 2-30 above in a power series in $\beta$, so that we may more easily compare them with the predictions of the classical theory of the Doppler effect and with experiment. Let us write Eq. 2-30a in the form

$$\nu = \nu_0 (1 + \beta)^{1/2} (1 - \beta)^{-1/2}$$

and expand each of the quantities in parentheses by the binomial theorem. We obtain

$$\nu = \nu_o (1 + \tfrac{1}{2}\beta - \tfrac{1}{8}\beta^2 + \cdots)(1 + \tfrac{1}{2}\beta^2 + \tfrac{3}{8}\beta + \cdots).$$

If we multiply this out, discarding terms of higher order than $\beta^2$, we find

$$\nu = \nu_0 (1 + \beta + \tfrac{1}{2}\beta^2 + \cdots) \quad \text{(approaching)}. \tag{2-31a}$$

Operating on Eq. 2-30b in the same way leads to

$$\nu = \nu_0 (1 - \beta + \tfrac{1}{2}\beta^2 - \cdots) \quad \text{(separating)}. \tag{2-31b}$$

It is instructive to compare these relativistic Doppler effect formulas with those derived from the classical ether or emission theories of the propagation of light (see *Physics*, Part II, Sec. 42-5). In each case, if the classical formulas are expanded in a power series in $\beta$, they agree with Eq. 2-31a (if source and observer are approaching) or with Eq. 2-31b (if source and observer are separating) *as far as the first two terms are concerned.*

The classical and the relativistic formulas differ in their predictions about the third term, that is, about the coefficients of $\beta^2$ in the power expansions, and it is at this level that comparisons between experiment and theory must be made. The experiments are difficult. If $\beta \ll 1$, then $\beta^2$ will be even smaller. Agreement between experiment and theory, clearly favoring the relativistic predictions (Eqs. 2-31) over the classical ones, was first obtained in 1938 by Ives and Stillwell (*Physics,* Part II, Sec. 42-5), following a suggestion first made by Einstein in 1907. They used a beam of excited hydrogen atoms of well-defined speed and direction as the source of the radiation whose Doppler frequency shift they studied. The experiment was repeated in 1961 with higher accuracy by Mandelberg and Witten [8], again confirming the relativistic predictions.

More striking, however, is the fact that the relativistic formula predicts a *transverse Doppler effect,* an effect that is purely relativistic, for there is no transverse Doppler effect in classical physics at all. This prediction follows from Eq. 2-27b when we set $\theta = 90°$, obtaining

$$\nu = \nu_0 \sqrt{1 - \beta^2} \quad \text{(transverse).} \qquad (2\text{-}32a)$$

If our line of sight is 90° to the motion of the source, then we should observe a frequency $\nu$ that is *lower* than the proper frequency $\nu_0$ of the source.

For easier comparison with the longitudinal Doppler effect (in the important case of $\beta \ll 1$), it is helpful to expand Eq. 2-32a as a power series in $\beta$, using the binomial theorem. Doing so yields

$$\nu = \nu_0(1 - \beta^2)^{1/2}$$

$$= \nu_0(1 - \tfrac{1}{2}\beta^2 + \ldots) \quad \text{(transverse).} \qquad (2\text{-}32b)$$

In comparison with Eqs. 2-31 the transverse Doppler formula contains no term in $\beta$. Recall that, in the formulas for the longitudinal Doppler effect (Eqs. 2-31), it is precisely this first-order term that we associate with the classical theory. Thus the absence of such a term in the transverse Doppler formula is totally consistent with the fact that classical theory does not predict such an effect.

Ives and Stillwell, in 1938 and 1941, were the first to confirm the existence of the transverse Doppler effect, using moving hydrogen atoms as "clocks." More recently, Walter Kundig [9] has obtained excellent quantitative data confirming the relativistic formula to within the experimental error of 1.1 percent. In Kundig's experiment a radioactive source emitting 14.4-keV gamma rays (for which the proper frequency is $3.48 \times 10^{18}$ Hz) was located on the axis of the rotor of a centrifuge. At the centrifuge rim was placed a resonant absorbing foil that is critically sensitive to the source frequency when the foil is at rest; a gamma-ray detector is placed behind the foil. When the centrifuge is operating, the absorbing foil is in rapid transverse motion with respect to the source and we expect that the characteristic absorption frequency of the foil will shift to a lower value, as predicted by the transverse Doppler formula (Eq. 2-32a). By sensitive Mössbauer techniques it is possible to change the effective frequency of the radiation emitted by the source and thus to measure this transverse Doppler frequency shift, as a function of rotor speed. Figure 2-12 shows that the experimental points fall very closely indeed on the curve predicted by relativity theory.

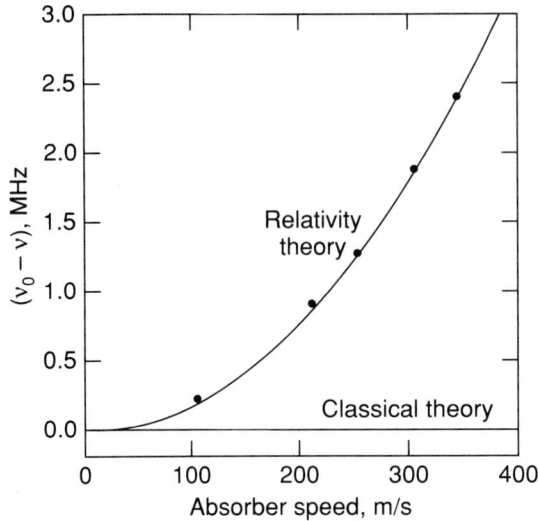

**FIGURE 2-12.**   The results of Kindig [9] on the transverse Doppler effect. The experimental points agree very well with the relativistic prediction and not at all with the classical prediction.

It is instructive to note that the transverse Doppler effect has a simple time-dilation interpretation. The moving source is really a moving clock, beating out electromagnetic oscillations. We have seen that moving clocks appear to run slow. Hence, we see a given number of oscillations in a time that is longer than the proper time. Or, equivalently, we see a smaller number of oscillations in our unit time than is seen in the unit time of the proper frame. Therefore, we observe a lower frequency than the proper frequency. The transverse Doppler effect is another physical example confirming the relativistic time dilation.

In both the Doppler effect and aberration, the theory of relativity introduces an intrinsic simplification over the classical interpretation of these effects in that the two separate cases, which are different in classical theory (namely, source at rest—moving observer and observer at rest—moving source), are identical in relativity. This, too, is in accord with observation. Notice, also, that a single derivation yields at once three effects, namely, aberration, longitudinal Doppler effect, and transverse Doppler effect.

**EXAMPLE 12.** _____

***Signaling from Space.*** *A*, on earth, signals with a flashlight every 6 min. *B* is on a space station that is stationary with respect to the earth. *C* is on a rocket traveling from *A* to *B* with a constant velocity of 0.6*c* relative to *A*; see Fig. 2-13. (*a*) At what intervals does *B* receive signals from *A*? (*b*) At what intervals does *C* receive signals from *A*? (*c*) If *C* flashes a light every time she receives

a flash from *A*, at what intervals does *B* receive *C*'s flashes?

(*a*) There is no relative motion of frames *A* and *B*, so, in effect, they are in the same inertial reference frame. Thus *B* receives signals from *A* every 6 min.

(*b*) Here the source (*A*) and the observer (*C*) are separating from each other. We use the lon-

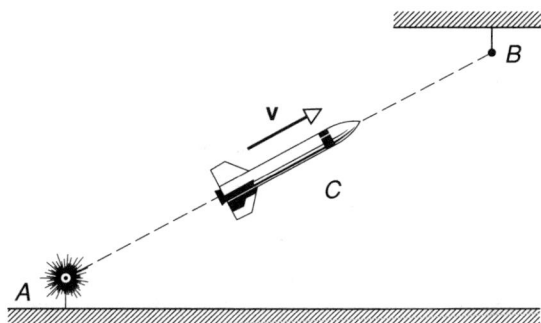

**FIGURE 2-13.** Example 12.

gitudinal Doppler effect formula, Eq. 2-30*b*, written as

$$\nu_C = \nu_A \sqrt{\frac{1 - \beta}{1 + \beta}},$$

where $\nu_C$ refers to the observer (rocket frame) and $\nu_A$ to the source (earth frame). Hence,

$$\nu_C = \nu_A \sqrt{\frac{1 - 0.6}{1 + 0.6}} = \frac{\nu_A}{2}.$$

But the period $T_A (= 1/\nu_A)$ equals 6 min. Therefore,

$$T_C = \frac{1}{\nu_C} = \frac{2}{\nu_A} = 2T_A = 12 \text{ min.}$$

Thus $C$ receives signals from $A$ at 12-min intervals.

(*c*) $C$ sends signals to $B$ at the same frequency $\nu_C$ that she receives them from $A$. Let $\nu_B$ be the frequency of the signals received by $B$ from $C$. Because $C$ (who is now the source) and $B$ (the observer) are *approaching* each other, we use the Doppler formula Eq. 2-30*a*, which gives us

$$\nu_B = \nu_C \sqrt{\frac{1 + \beta}{1 - \beta}}$$

$$= \left( \nu_A \sqrt{\frac{1 - \beta}{1 + \beta}} \right) \sqrt{\frac{1 + \beta}{1 - \beta}} = \nu_A.$$

Since $\nu_B = \nu_A$, it follows that $T_B = T_A$ and thus $B$ receives signals from $C$ at 6-min intervals. This is the same rate at which $B$ receives signals directly from $A$. Explain why this is plausible.

## 2-8 The Common Sense of Special Relativity

We are now at a point where a retrospective view can be helpful. Special relativity theory makes still more predictions than we have discussed so far that contradict classical views. Later we shall see that, in those cases too, experimental results confirm the relativistic predictions. Indeed, in many branches of physics, whether the subject is elementary particles, nuclei, atoms, stars, or the universe itself, relativity is used in an almost commonplace way as the correct description of the real microscopic world. Furthermore, relativity is a consistent theory, as we have shown already in many ways and shall continue to show later. However, because our everyday macroscopic world is classical to a good approximation and students have not yet lived with or used relativity enough to become sufficiently familiar with it, there may remain misconceptions about the theory that are worth discussing now.

**The Limiting Speed c of Signals** We have seen that, if it were possible to transmit signals with infinite speed, we could establish in an absolute way whether or not two events are simultaneous. The relativity of simultaneity depended on the existence of a finite speed of transmission of signals. Now, we probably would grant that it is unrealistic to expect that any physical action could be transmitted with infinite speed. It does indeed seem fanciful that we could initiate a signal that would travel to all parts of our universe in zero time. It is really the classical physics (which at bottom makes such an assumption) that is fictitious (science fiction) and not the relativistic physics, which postulates a limiting speed. Furthermore, when experiments are carried out, the relativity of time measurements is confirmed.

Nature does indeed show that relativity is a practical theory of measurement and not a philosophically idealistic one, as is the classical theory.

We can look at this in another way. From the fact that experiment denies the absolute nature of time, we can conclude that signals cannot be transmitted with infinite speed. Hence, there must be a certain finite speed that cannot be exceeded and that we call the limiting speed. The principle of relativity shows at once that this limiting speed is the speed of light, since the result that no speed can exceed a given limit is certainly a law of physics and, according to the principle of relativity, the laws of physics are the same for all inertial observers. Therefore this given limit, the limiting speed, must be exactly the same in all inertial reference frames. We have seen, from experiment, that the speed of light has exactly this property.

Viewed in this way, the speed of electromagnetic waves in a vacuum assumes a role wider than the travel rate of a particular physical entity. It becomes instead a limiting speed for the motion of anything in nature.

**Absolutism and Relativity**   The theory of relativity could have been called the theory of absolutism, with some justification. The fact that the observers who are in relative motion assign different numbers to length and time intervals between the pair of events, rather than finding these numbers to be absolutes, upsets the classical mind. This is so in spite of the fact that even in classical physics the measured values of the momentum or kinetic energy of a particle, for example, also are different for two observers who are in relative motion. What is troublesome, apparently, is the philosophic notion that length and time in the abstract are absolute quantities and the belief that relativity contradicts this notion. Now, without going into such a philosophic byway, it is important to note that relativity simply says that the *measured* length or time interval between a pair of events is affected by the relative *motion* of the events and measurer. Relativity is a theory of measurement, and motion affects measurement. Let us look at various aspects of this.

That relative motion should affect measurements is almost a "common-sense" idea—classical physics is full of such examples, including the aberration and Doppler effects already discussed. Furthermore, to explain such phenomena in relativity, we need not talk about the structure of matter or the idea of an ether in order to find changes in length and duration due to motion. Instead, the results follow directly *from the measurement process itself*. Indeed, we find that the phenomena are *reciprocal*. That is, just as A's clock seems to B to run slow, so does B's clock seem to run slow to A; just as A's meter stick seems to B to have contracted in the direction of motion, so likewise B's meter stick seems to A to have contracted in exactly the same way.

Moreover, we should note that in a narrower sense there *are* absolute lengths and times in relativity. The *rest length* of a rod is an absolute quantity, the same for all inertial observers: If a given rod is measured by different inertial observers by bringing the rod to rest in their respective frames, each will measure the same length. Similarly for clocks, the *proper time* (which might better have been called "local time") is an absolute quantity: The frequency of oscillation of an ammonia molecule, for instance, would be measured to be the same by different inertial observers who bring the molecule to rest in their respective frames.

The separation in space of two events (length of a rod) and the time interval between them (rate of a clock) are absolute quantities in classical (Galilean) physics, even for observers in relative motion. At first glance it may seem a step backward to learn that these quantities have surrendered their absolute character in relativity theory in that they have different values for different inertial observers. We learned in Section 2-3, however, that relativity has given us a broader perspective in this matter by replacing two separately absolute quantities by a single absolute, the *spacetime interval,* which has the same value for all observers. The result is a new understanding of the nature of space and time, an understanding that is at the same time both simpler and more profound.

Where relativity theory is clearly ''more absolute'' than classical physics is in the relativity principle itself: The *laws of physics* are absolute. We have seen that the Galilean transformations and classical notions contradicted the invariance of electromagnetic (and optical) laws, for example. Surely, giving up the absoluteness of the laws of physics, as classical notions of time and length demand, would leave us with an arbitrary and complex physical world. By comparison, relativity is absolute and simple.

**The Reality of the Length Contraction**  Is the length contraction ''real'' or apparent? We might answer this by posing a similar question. Is the frequency, or wavelength, shift in the Doppler effect real or apparent? Certainly the proper frequency (that is, the rest frequency) of the source is measured to be the same by all observers who bring the source to rest before taking the measurement. Likewise, the proper length is invariant. When the source and observer are in relative motion, the observer definitely measures a frequency (or wavelength) shift. Likewise, the moving rod is definitely measured to be contracted. The effects are real in the same sense that the measurements are real. We do not claim that the proper frequency has changed because of our measured shift. Nor do we claim that the proper length has changed because of our measured contraction. The effects are apparent (that is, caused by the motion) in the same sense that proper quantities have not changed.

We do not speak about theories of matter to explain the contraction but, instead, we invoke the measurement process itself. For example, we do not assert, as Lorentz sought to prove, that motion produces a physical contraction through an effect on the elastic forces in the electronic or atomic constitution of matter (motion is *relative,* not absolute), but instead we remember the fish story. If a fish is swimming in water and his length is the distance between his tail and his nose, measured simultaneously, observers who disagree on whether measurements are simultaneous or not will certainly disagree on the measured length. Hence, length contraction is due to the relativity of simultaneity.

Since length measurements involve a comparison of two lengths (moving rod and measuring rod, for example), we can see that the Lorentz length contraction is really not a property of a single rod by itself but instead is a relation between two such rods in relative motion. The relation is both observable and reciprocal. Just as A's meter stick seems to B to have contracted in the direction of motion, so likewise B's meter stick seems to A to have contracted, in exactly the same way.

**Rigid Bodies and Unit Length** In classical physics, the notion of an ideal rigid body was often used as the basis for length (that is, space) measurements. In principle, a rigid rod of unit length is used to lay out a distance scale. Even in relativity we can imagine a standard rod defining a unit distance, this same rod being brought to rest in each observer's frame to lay out space-coordinate units. However, the concept of an ideal rigid body is untenable in relativity, for such a body would be capable of transmitting signals instantaneously; a disturbance at one end would be propagated with infinite velocity through the body, in contradiction to the relativistic principle that there is a finite upper limit to the speed of transmission of a signal.

Conceptually, then, we must give up the notion of an ideal rigid body. This causes no problems because time measurements prove to be primary and space measurements secondary. We know that this is so in relativity, for the simultaneity concept is used in the definition of length. But a similar situation also exists in classical physics. Some years have passed since distances were measured in terms of comparison with a presumed rigid measuring rod, the standard meter. This definition was replaced in 1960 by a definition of the meter in terms of the wavelength of the radiation emitted by a specified light source. Since 1983 the meter has been defined as "the length equal to the distance traveled in a time interval of 1/299,792,458 of a second by plane electromagnetic waves in a vacuum." Thus length is now measured by timing a light beam, the speed of light having a *defined* value of 299,792,458 m/s. It is interesting that this definition of the meter, arrived at after careful consideration by a representative international body, contains no mention of either a standard frequency or a standard reference frame. Those making the choice evidently took it for granted (following Maxwell) that the speed of electromagnetic radiation is independent of frequency and (following Einstein) that it is the same for all inertial observers.

Rigid-body measuring concepts have never been directly applicable in certain situations, measurements on the atomic and the astronomical scales being two limiting examples. As the unit "light-year" suggests, the timing concept that now forms the official basis of all length measurement, has served as the basis of much practical distance measurement, the precise measurement of the earth–moon distance by radar techniques being one example. It seems clear that rigid-body measuring rods—not permissible conceptually in relativity theory—are not required in practice, even for measurements in the classical realm.

It is fitting, in emphasizing the common sense of relativity, to conclude with this quotation from Bondi [10] on the presentation of relativity theory:

> At first, relativity was considered shocking, anti-establishment and highly mysterious, and all presentations intended for the population at large were meant to emphasize these shocking and mysterious aspects, which is hardly conducive to easy teaching and good understanding. They tended to emphasize the revolutionary aspects of the theory whereas, surely, it would be good teaching to emphasize the continuity with earlier thought. . . .
>
> It is first necessary to bring home to the student very clearly the Newtonian attitude. Newton's first law of dynamics leads directly to the notion of an *inertial observer,* defined as an observer who finds the law of inertia to be correct. . . . The utter equivalence of inertial observers to each other for the purpose of Newton's first law is

a direct and logical consequence of this law. The equivalence with regard to the second law is not a logical necessity but a very plausible extension, and with this plausible extension we arrive at Newton's principle of relativity: *that all inertial observers are equivalent as far as dynamical experiments go*. It will be obvious that the restriction to dynamical experiments is due simply to this principle of relativity having been derived from the laws of dynamics. . . .

The next step . . . is to point out how absurd it would be if dynamics were in any sense separated from the rest of physics. There is no experiment in physics that involves dynamics alone and nothing else. . . . Hence, Newton's principle of relativity is empty because it refers only to a class of experiment that does not exist—the purely dynamical experiment. The choice is therefore presented of either throwing out this principle or removing its restriction to dynamical experiments. The first alternative does not lead us any further, and clearly disregards something of significance in our experience. The second alternative immediately gives us Einstein's principle of relativity: *that all inertial observers are equivalent*. It presents this principle, not as a logical deduction, but as a reasonable guess, a fertile guess from which observable consequences may be derived so that this particular hypothesis can be subjected to experimental testing. Thus, the principle of relativity is seen, not as a revolutionary new step, but as a natural, indeed as an almost obvious, completion of Newton's work.

# QUESTIONS

1. Distinguish between sound and light as to their value as synchronizing signals. Is there a lack of analogy?

2. Give an example from classical physics in which the motion of a clock affects its rate, that is, the way it runs. (The magnitude of the effect may depend on the detailed nature of the clock.)

3. Explain how the result of the Michelson-Morley experiment was put into our definition (procedure) of simultaneity (for synchronizing clocks).

4. According to Eqs. 2-4 and 2-5, each inertial observer finds the center of the spherical electromagnetic wave to be at his own origin at all times, even when the origins do not coincide. How is this result related to our procedure for synchronizing clocks?

5. What assumptions, other than the relativity principle and the principle of the constancy of the speed of light, were made in deducing the Lorentz transformation equations?

6. Two observers, one at rest in $S$ and one at rest in $S'$, each carry a meter stick oriented parallel to their relative motion. *Each* observer finds upon measurement that the *other* observer's meter stick is shorter than his own meter stick. Does this seem like a paradox to you? Explain. (*Hint:* Compare the following situation. Harry waves goodbye to Walter who is in the rear of a station wagon driving away from Harry. Harry says that Walter gets smaller. Walter says that Harry gets smaller. Are they measuring the same thing?)

7. Although in relativity (where motion is relative and not absolute) we find that "moving clocks run slow," this effect has nothing to do with the motion altering the way a clock works. What does it have to do with?

8. Two events occur at the same place and at the same time for one observer. Will they be simultaneous for all other observers? Will they also occur at the same place for all other observers?

9. Events *A* and *B* occur at the same point in a certain inertial reference frame, with event *A* preceding event *B*. Will *A* precede *B* in all other frames? In any other frame? Will the events occur at the same point in any other frame? Will the time interval between the events be the same in any other frame?

10. Two events are simultaneous but separated in space in one inertial reference frame. Will they be simultaneous in any other frame? Will their spatial separation be the same in any other frame?

11. We have seen that if several observers watch two events, labeled *A* and *B*, one of them may say that event *A* occurred first but another may claim that it was event *B* that did so. What would you say to a friend who asked you which event *really did* occur first?

12. Let event *A* be the departure of an airplane from San Francisco and event *B* be its arrival in New York. Is it possible to find two observers who disagree about the time order of *these* events? Explain.

13. A rod has a tiny flashbulb embedded in each end. Each of these bulbs has been arranged to flash independently, at a single unpredictable time. The flashes enable an observer to measure the coordinates of the end points of the rod in his reference frame. Consider first an observer in frame *S*, in which the rod is at rest and lying along the *x* axis. Would you label the difference between his coordinate readings as the ''length of the rod''? As the ''rest length of the rod''? Answer these questions for a observer *S'* with respect to whom the rod is moving at speed *v*.

14. A number of observers, in different inertial reference frames, measure the spacetime coordinates of two events. For what combination(s) of these measurements would all of these observers obtain the same numerical value?

15. How would you recognize a given pair of events as spacelike? Timelike? Lightlike?

16. Can a given pair of events appear spacelike to one inertial observer and timelike to another?

17. Explain, using the velocity addition theorem of relativity, how we can account for the result of the Michelson-Morley experiment and for the double-star observations.

18. In Example 10, what would happen if $v_w$ were chosen equal to $-c/n$?

19. The Galilean velocity transformation equation (Eq. 2-19) is so instinctively familiar from everyday experience that it is sometimes claimed to be ''obviously correct, requiring no proof.'' Many so-called refutations of relativity theory turn out to be based on this claim. How would you refute someone who made such a claim?

20. Compare the results obtained for length- and time-interval measurements by observers in reference frames whose relative velocity is *c*. In what sense, from

this point of view, does $c$ appear as a limiting velocity, to be approached (but not attained) by material bodies?

21. Is the Doppler effect simply a time-dilation effect and nothing more, or is there something else to it?

22. An observer makes simultaneous measurements of the positions of the end points of a rod that is lying at rest along the $x$ axis of his reference frame. An observer in another reference frame who views these same measuring events will always find (can you prove it?) that the difference between his position measurements is *greater* than the rest length of the rod. How can you square this with the fact that moving rods have measured lengths that are *less than* their rest lengths?

23. Consider a spherical light wavefront spreading out from a point source. As seen by an observer at the source, what is the *difference in velocity* of portions of the wavefront traveling in opposite directions? What is the *relative velocity* of one of these portions of the wavefront with respect to the other?

24. The sweep rate of the tail of a comet can exceed the speed of light. Explain this phenomenon and show that there is no contradiction with relativity.

25. Explain, in qualitative terms, the "headlight effect" described in Problem 82.

26. List several experimental results not predicted or explained by classical physics that are predicted or explained by the theory of relativity.

27. We have stressed the utility of relativity at high speeds. Relativity is also useful in cosmology, where great distances and large time intervals are involved. Show, from the form of the Lorentz transformation equations, why this is so.

28. In relativity the time and space coordinates are intertwined and treated on a more or less equivalent basis. Are time and space fundamentally of the same nature, or is there some essential difference between them, preserved even in relativity? (*Hint:* What significance do you attach to the minus sign in Eq. 2-16?)

29. We have stressed certain measurements of events in spacetime that are different for different inertial observers. Make a list of those things that these different observers *agree* on.

30. Some say that relativity complicates things. Give examples to the contrary, wherein relativity simplifies matters. Consider the Fresnel drag experiment as one example.

# PROBLEMS

*In all problems that involve two reference frames it is assumed, unless otherwise stated, that the frames are in the standard configuration of Fig. 1-1. That is, an observer in the S frame would see the S' frame moving with speed $v$ in the direction of increasing x. The x and x' axes coincide and the y-y' and z-z' axes remain parallel. Further, it is assumed that each observer sets his or her clocks to zero at the instant the two origins pass each other.*

1. **The view from the other frame.**   In Fig. 2-1 we took the point of view of observer $O$ in the $S$ frame and found that events $AA'$ and $BB'$ happened to be simultaneous in that frame. Figure 2-14 shows how these *same two events* appear to observer $O'$ in the $S'$ frame. Note, and comment on, the following features: (*a*) $O'$ is midway between points $A'$ and $B'$. (*b*) $S$ moves to the left with speed $v$. (*c*) The first stroke occurs on the right, making marks $B$, $B'$. (*d*) Later, a stroke occurs on the left, making marks $A$, $A'$. (*e*) As seen by $O'$, the distances of $AB$ and $A'B'$ are *necessarily* unequal. (*f*) In agreement with the observations of observer $O$ (for these same events) described in connection with Fig. 2-1, the right wavefront passes $O'$, then both wavefronts pass $O$, and finally the left wavefront passes $O'$. (*g*) How does the situation described in this figure differ from that described in Fig. 2-2, which *also* purports to show the point of view of the $S'$ observer?

2. **A little algebra.**   Show that Eqs. 2-6 for $a_{11}$, $a_{41}$, and $a_{44}$ are indeed solutions of the equations preceeding them.

3. **Three simplifying physical arguments.** Consider the two middle equations of Eqs. 2-1:

$$y' = a_{21}x + a_{22}y + a_{23}z + a_{24}t$$
$$z' = a_{31}x + a_{32}y + a_{33}z + a_{34}t.$$

(*a*) As part of the derivation of the Lorentz transformation equations given in Section 2-2, the $S$ and $S'$ frames have been so arranged that their $x$ and $x'$ axes coincide, as in Fig. 2-1. Show how to deduce from this fact that

$$a_{21} = a_{24} = a_{31} = a_{34} = 0.$$

(*b*) From the fact that the $y$ and $y'$ axes (and the $z$ and $z'$ axes) have been taken to be parallel, show that

$$a_{23} = a_{32} = 0.$$

(*c*) Let observer $S$ place a rod, whose rest length is 1 m, along her $y$ axis and let observer $S'$ measure its length as it moves past. Then let $S'$ place the same rod along the $y'$ axis and let $S$ measure it. By applying the principle of relativity to their results, show that we must have

$$a_{22} = a_{33} = 1.$$

(*d*) With these substitutions, to what do the expressions given above for $y'$ and $z'$ reduce?

4. **A feeling for the Lorentz factor.**   The Lorentz factor $\gamma$ (see Eq. 2-10) is a direct measure of, among other quantities, the relativistic length contraction and the time-dilation effect; see Eqs. 2-12*a* and 2.14*b*. What must be the relative speed parameter $\beta$ of two reference frames if this factor is to be (*a*) 1.01? (*b*) 10? (*c*) 100? (*d*) 1000?

5. **Making absolutely sure.**   In Table 2-2 the Lorentz transformation equations in the right-hand column can be derived from those

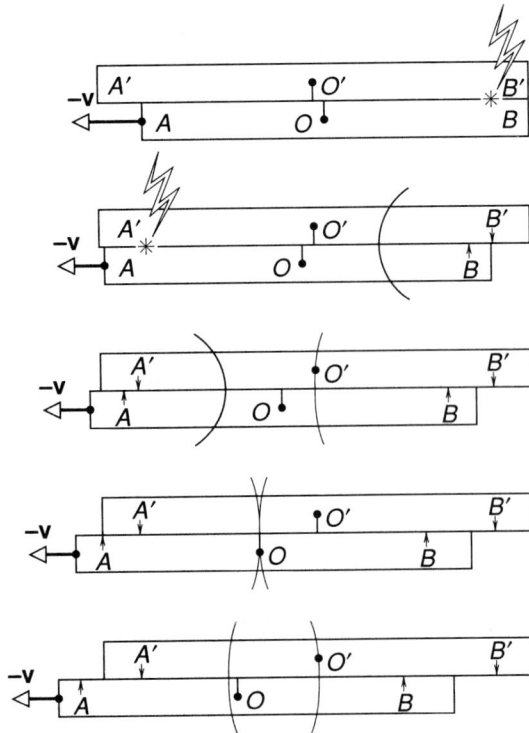

**FIGURE 2-14.**   Problem 1.

in the left-hand column simply by (1) exchanging primed and unprimed quantities and (2) changing the sign of $v$. Verify this procedure by deriving one set of equations directly from the other by algebraic manipulation.

6. **The speed of light really is the same in all frames.** Equation 2-4 describes an expanding spherical wavefront of light, triggered at $t = 0$ and viewed in the $S$ frame. Equation 2-5 describes the same expanding wavefront as viewed in the $S'$ frame. Show that either equation can be derived from the other by direct application of the Lorentz transformation equations.

7. **Two observers view the same event (I).** Observer $S$ assigns the following spacetime coordinates to an event:

$$x = 100 \text{ km} \qquad y = 10 \text{ km}$$
$$z = 55 \text{ km} \qquad t = 200 \text{ } \mu s.$$

What are the coordinates of this event in frame $S'$, which moves in the direction of increasing $x$ with speed $0.95c$? Check your answers by using the inverse Lorentz transformation equations to obtain the original data.

8. **Two observers view the same event (II).** Observer $S$ reports that an event occurred on his $x$ axis at $x = 3.0 \times 10^8$ $m$ at a time $t = 2.50$ s. (a) Observer $S'$ is moving in the direction of increasing $x$ at a speed of $0.40c$. What coordinates would he report for the event? (b) What coordinates would he report if he were moving in the direction of *decreasing* $x$ at this same speed?

9. **A moving clock.** A clock moves along the $x$ axis at a speed of $0.60c$ and reads zero as it passes the origin. What time does it read as it passes the 180-m mark on this axis?

10. **A moving rod.** A rod lies parallel to the $x$ axis of reference frame $S$, moving along this axis at a speed of $0.60c$. Its rest length is 1.0 m. What will be its measured length in frame $S$?

11. **Hidden symmetry in the Lorentz equations.** The spacetime symmetry of relativity is partially concealed because the space variables and the time variable are measured in different units. Let us introduce a new variable $w$ ($= ct$) that is a linear measure of time but is expressed in length units. (a) Show that the first and fourth Lorentz transformation equations can be written in the forms

$$x' = \gamma(x - \beta w) \qquad x = \gamma(x' + \beta w')$$
$$w' = \gamma(w - \beta x) \qquad w = \gamma(w' + \beta x'),$$

which are certainly more symmetrical in appearance than the standard representation of Table 2-2. (b) The variable $w$ can be expressed in "meters of time." Show that, with this understanding, 1 $\mu s \equiv 300$ m. Use this representation of the Lorentz transformation equations to solve Problem 7.

12. **Meters of time.** In a frame $S$, two events occur along the $x$ axis. They are separated in space by $\Delta x$ ($= x_2 - x_1 = 720$ m) and in the time coordinate (see Problem 11) by $\Delta w$ ($= w_2 - w_1 = 1500$ m). (a) What is the relative speed parameter $\beta$ of a second frame $S'$ in which these events are found to occur at the same place? (b) What is the time interval between them in this second frame? Recall from Problem 11 that 1 $\mu s$ is equivalent to 300 "meters of time."

13. **A fast spaceship.** The length of a spaceship is measured to be exactly half its rest length. (a) What is the speed of the spaceship relative to the observer's frame? (b) By what factor do the spaceship's clocks run slow, compared to clocks in the observer's frame?

14. **A slow airplane.** An airplane whose rest length is 40.0 m is moving at a uniform velocity with respect to the earth at a speed of 630 m/s. (a) By what fraction of its rest length will it appear to be shortened to an observer on earth? (b) How long would it take by earth clocks for the airplane's clock to fall behind by 1 $\mu s$? (Assume that only special relativity applies.)

15. **Riding a fast electron.** A 100-MeV electron, for which $\beta = 0.999987$, moves along the axis of an evacuated tube that has a length of 3.00 m as measured by a laboratory observer $S$ with respect to whom the tube is at rest. An observer $S'$ moving with the electron, however, would see this tube moving past her with speed $v\ (= \beta c)$. What length would this observer measure for the tube? (*Hint:* See Example 3*b*.)

16. **Watching the earth drift by.** The rest radius of the earth is 6400 km and its orbital speed about the sun is 30 km/s. By how much would the earth's diameter appear to be shortened to an observer stationed so as to be able to watch the earth move past him at this speed?

17. **A world that never was.** Consider a universe in which the speed of light $c$ is equal to 100 mi/h. A Lincoln Continental traveling at a speed $v$ relative to a fixed radar speed trap overtakes a Volkswagen traveling at the speed limit of 50 mi/h. The Lincoln's speed is such that its length, as measured by the fixed observer, is the same as that of the Volkswagen. By how much is the Lincoln exceeding the speed limit? At rest, a Lincoln is twice as long as a Volkswagen.

18. **A moving, slanting, rod.** A thin rod of length $L'$, at rest in the $S'$ frame, makes an angle of $\theta'$ with the $x'$ axis, as in Fig. 2-15.

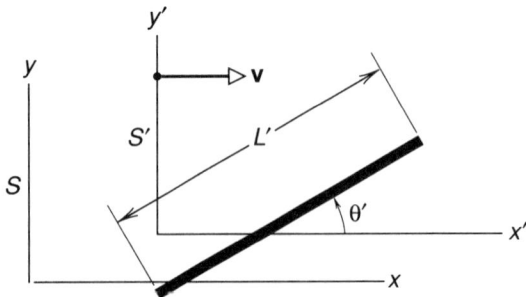

**FIGURE 2-15.** Problem 18.

(*a*) What is its length $L$ as measured by an observer in the $S$ frame, for whom the rod is

moving at a speed of $\beta c$ in the direction of increasing $x$? (*b*) What angle $\theta$ does this moving rod make with the $x$ axis? (*c*) Evaluate these quantities for $L' = 1.00$ m, $\theta' = 30°$, and $\beta = 0.40$.

19. **Timing a spaceship.** A spaceship of rest length 100 m drifts past a timing station at a speed of 0.80$c$. What time interval between the passage of the front and the back end of the ship will the station monitor record?

20. **A journey to Vega.** A space traveler takes off from earth and moves at speed 0.990$c$ toward the star Vega, which is 26 ly distant. How much time will have elapsed by earth clocks (*a*) When the traveler reaches Vega? (*b*) When the earth observers receive word from him that he has arrived? (*c*) How much older will the earth observers calculate the traveler to be when he reaches Vega than he was when he started the trip?

21. **To the galactic center!** (*a*) Can a person, in principle, travel from earth to the galactic center (which is about 28,000 ly distant) in a normal lifetime? Explain, using either time-dilation or length-contraction arguments. (*b*) What constant velocity would be needed to make the trip in 30 y (proper time)?

22. **Pretty small but quite measureable.** An airline pilot synchronizes his watch with earth clocks and then takes off on a nonstop flight of 6000 mi at a steady speed of 600 mi/h. On landing, how far behind would his watch be compared to earth clocks, assuming that special relativity alone applies? (An around-the-world flight using atomic clocks confirmed this effect, as well as a separate general-relativistic effect associated with the earth's gravitational field [11]).

23. **Decay in flight (I).** A pion is created in the higher reaches of the earth's atmosphere when an incoming high-energy cosmic-ray particle collides with an atomic nucleus. A pion so formed descends toward earth with a speed of 0.990$c$. In a reference frame in which they are at rest, pions decay with a

mean life of 26 ns. As measured in a frame fixed with respect to the earth, how far (on the average) will such a typical pion move through the atmosphere before it decays?

**24. Decay in flight (II).** The mean lifetime of muons stopped in a lead block in the laboratory is measured to be 2.2 μs. The mean lifetime of high-speed muons in a burst of cosmic rays observed from the earth is measured to be 16 μs. Find the speed of these cosmic ray muons.

**25. Decay in flight (III).** An unstable high-energy particle enters a detector and leaves a track 1.05 mm long before it decays. Its speed relative to the detector was $0.992c$. What is its proper lifetime? That is, how long would it have lasted before decay had it been at rest with respect to the detector?

**26. Decay in flight (IV).** In the target area of an accelerator laboratory there is a straight evacuated tube 300 m long. A momentary burst of 1 million radioactive particles enters at one end of the tube, moving at a speed of $0.80c$. Half of them arrive at the other end without having decayed. (*a*) How long is the tube as measured by an observer moving with the particles? (*b*) What is the half-life of the particles (that is, the time during which half of the particles initially present have decayed) in this same reference frame? (*c*) With what speed is the tube measured to move in this frame?

**27. Decay in flight (V).** (*a*) If the average (proper) lifetime of a muon is 2.2 μs, what average distance would it travel in free space before decaying, as measured in reference frames in which its velocity is $0.00c$; $0.60c$; $0.90c$; $0.990c$? (*b*) Compare each of these distances with the distance the muon sees itself traveling through.

**28. Simultaneous—but to whom?** An experimenter arranges to trigger two flashbulbs simultaneously, a blue flash located at the origin of his reference frame and a red flash at $x = 30$ km. A second observer, moving at a speed of $0.25c$ in the direction of increasing $x$, also views these flashes. (*a*) What time interval between them does he find? (*b*) Which flash does he say occurs first?

**29. Simultaneity—the general case.** Two events, one at position $x_1$, $y_1$, $z_1$ and another at a different position $x_2$, $y_2$, $z_2$ occur at the same time according to observer $S$. (*a*) Do these events appear to be simultaneous to observer $S'$, who moves relative to $S$ at speed $v$? (*b*) If not, what is the time interval that $S'$ measures between these events? (*c*) How is this interval affected as $v \to 0$? As $v \to c$? As the separation between the events goes to zero?

**30. A string of lights across the desert.** Observer $S$ sees a series of light flashes extending indefinitely in a straight line across a desert. By measurement he declares them all to have occurred simultaneously and finds further that adjacent flashes were uniformly separated by 3.0 km. Observer $S'$ is moving along this line with a speed $0.50c$. What are the results of *his* measurements of the space-time coordinates of these light flashes?

**31. Careful measurements on a moving flat-car (I).** A flatcar moves on a track at a constant speed $v$ (see Fig. 2-16). Observers $A$ and $B$ are on the ends of the car and observers $C$ and $D$ are stationed along the track. We define event $AC$ as the occurrence of $A$ passing $C$, and the others similarly. (*a*) Of the four events $BD$, $BC$, $AD$, $AC$, which are useful for observers along the track who wish to measure the rate of a clock carried by $A$? (*b*) Let $\Delta t$ be the time interval between these two events for the track observers. What time interval does the moving clock show? (*c*) Suppose that events $BC$ and $AD$ are simultaneous to the track observers. Are they simultaneous to the observers on the flatcar? If not, which event is earlier?

**32. Careful measurements on a moving flat-car (II).** In Problem 31 (see Fig. 2-16), event $AD$ turns out to be simultaneous with

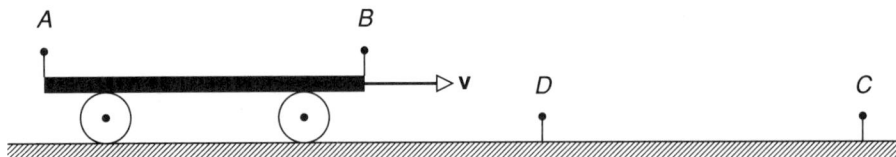

**FIGURE 2-16.**   Problems 31 and 32.

*BC* in the track frame. (*a*) The track observers set out to measure *AB,* the length of the car. They can do so either by using the events *BD* and *AD* and working through time measurements or by using events *BC* and *AC.* In either case, the car observers are not apt to regard these results as valid. Explain why for each case. (*b*) Suppose that the car observers seek to measure the distance *DC* by making simultaneous marks on a long meter stick. Where (relative to *A* and *B*) would an observer *E* have to be situated such that *AD* is simultaneous with *EC* in the car frame? Explain why in terms of synchronization. Can you see why there is a length contraction?

**33. *S* and *S'* time two events.**   Inertial frame *S'* moves at a speed of 0.60*c* with respect to frame *S*. Two events are recorded. In frame *S*, event 1 occurs at the origin at *t* = 0 and event 2 occurs on the *x* axis at *x* = 3.0 km and at *t* = 4.0 μs. What time of occurrence does observer *S'* record for these same events? Explain the difference in the time order.

**34. Two flashes at different places—or are they?**   An observer *S* sees a flash of red light 1200 m from his position and a flash of blue light 720 m closer to him and on the same straight line. He measures the time interval between the occurrence of the flashes to be 5.00 μs, the red flash occurring first. (*a*) What is the relative velocity **v** (magnitude and direction) of a second observer *S'* who would record these flashes as occurring at the same place? (*b*) From the point of view of *S'*, which flash occurs first? (*c*) What time interval between them would *S'* measure?

**35. The limit of possibility.**   In Problem 34, observer *S* sees the two flashes in the same positions, but they now occur closer together in time. How close together in time can they be and still have it possible to find a frame *S'* in which they occur at the same place?

**★36. What time is it anyway?**   Observers *S* and *S'* stand at the origins of their respective frames, which are moving relative to each other with a speed of 0.60*c*. Each has a standard clock, which, as usual, they set to zero when the two origins coincide. Observer *S* keeps the *S'* clock visually in sight. (*a*) What time will the *S'* clock record when the *S* clock records 5.00 μs? (*b*) What time will observer *S actually read* on the *S'* clock when his own clock reads 5.00 μs?

**37. A long train struck curiously by lightning.**   Assume, in Fig. 2-1, that *S'* is a train having a speed of 100 mi/h and that it is 0.50 mi long (rest length). What is the elapsed time between the arrival of the two wavefronts at *O'?* Do this in two ways: (*a*) Make the discussion we had in connection with Fig. 2-1 quantitative by finding expressions for the arrival times of the two signals at *O'* and subtracting them. Bear in mind that, from the point of view of *S'*, the distance between the marks at *A* and *B* in *S* is contracted by the Lorentz factor γ. (*b*) Treat the formation of the marks at *A* and *B* in frame *S* as two "events" and apply the Lorentz transformation equations to find their time separation in frame *S'*. To what prediction does your expression reduce if the train is at rest on the tracks? . . . if the speed of light suddenly becomes infinitely great?

**38. A comforting thought.** An observer at rest in the laboratory ($S$ frame) sees a uranium nucleus at the origin of this frame emit an alpha particle at time $t = 0$. The alpha particle travels along the $+x$ axis to position $x_1$, where, at time $t_1$, it is absorbed by a radon nucleus, to form an atom of radium. Show that a second observer in frame $S'$, moving along the $x$-$x'$ axis at speed $v$, cannot see the alpha particle absorbed by the radon nucleus before it is emitted by the uranium nucleus. Assume that all speeds involved are less than the speed of light.

**39. Two events—same time difference, same separation, different sequence.** Observer $S$ notes that two colored flashes of light, separated by 2400 m, occur along the positive branch of the $x$ axis of his reference frame. A blue flash occurs first, followed after 5.00 μs by a red flash, the latter being the most distant from the origin of his reference frame. A second observer $S'$ obtains exactly the same numerical values for both the time difference and the absolute spatial separation between the two events but declares that the *red* flash occurs first. (*a*) What is the relative speed of $S'$ with respect to $S$? (*b*) Which flash will $S'$ find to be the more distant from the origin of *her* reference frame?

**40. Can it be done?** In Problem 39, assume no other change but that observer $S$ determines that the blue flash occurs first. In particular, the spatial order of the two flashes is to remain unchanged, the red flash still being farther from the $S$ origin. Is it still possible to find a frame $S'$ in which the time interval between the events would remain unchanged but the order of events would be reversed?

**41. The interval is invariant—prove it!** Show that the (square of) the spacetime interval $(\Delta s)^2$ associated with two events (assumed to occur on the $x$-$x'$ axis) is invariant under a Lorentz transformation. That is (see Eq. 2-16), show that

$$(c\Delta t)^2 - (\Delta x)^2 = (c\Delta t')^2 - (\Delta x')^2.$$

(*Hint:* Use Table 2-3.)

**42. The interval is invariant—check it out!** Two events occur on the $x$ axis of reference frame $S$, their spacetime coordinates being:

| Event | $x$ | $t$ |
|-------|-----|-----|
| 1 | 720 m | 5.0 μs |
| 2 | 1200 m | 2.0 μs |

(*a*) What is the square of the spacetime interval $(\Delta s)^2$ for these two events? (Recall that the speed of light can be written as 300 m/μs.) (*b*) What are the coordinates of these events in a frame $S'$ that moves at speed $0.60c$ in the direction of increasing $x$? Calculate the square of the interval in this frame and compare it to the value calculated for frame $S$. (*c*) What are the coordinates of these events in a frame $S''$ that moves at speed $0.95c$ in the direction of decreasing $x$? Again, calculate $(\Delta s)^2$ and compare its value with the values found in (*a*) and (*b*). Do your calculations bear out the invariance of the spacetime interval?

**43. An event pair—timelike or spacelike?** Two events occur on the $x$ axis of reference frame $S$, their spacetime coordinates being:

| Event | $x$ | $t$ |
|-------|-----|-----|
| 1 | 200 m | 5.0 μs |
| 2 | 1200 m | 2.0 μs |

(*a*) What is the square of the spacetime interval $(\Delta s)^2$ for these two events? (*b*) What is the *proper distance* interval $\Delta\sigma$ between them? (*c*) If two events possess a (mathematically real) proper distance interval, it should be possible to find a frame $S'$ in which these events would be seen to occur simultaneously. Find this frame. (*d*) Can you calculate a (mathematically real) *proper time* interval $\Delta\tau$ for this pair of events? (e) Would you describe this pair of events as timelike? Spacelike? Lightlike? (Compare this problem carefully with Problem 42, noting that $(\Delta s)^2$ in that problem is positive and here it turns out to be negative.)

**44. An event pair—spacelike or timelike?**   In Problem 42 the spacetime coordinates of two x-axis events are given and three reference frames from which they might be viewed are described. (a) Using data from the solution to that problem, calculate the *proper time* interval $\Delta\tau$ for this pair of events, from the point of view of each of the three frames. Do your calculations support the claim that the proper time interval is an invariant quantity? The proper time interval that you have calculated should be *smaller* than any of the actual time intervals in the three given frames. Is it? (b) If two events have a (mathematically real) proper time interval between them, it ought to be possible to find a reference frame in which these events would be seen to occur at the same place. Find this frame. (c) Can you calculate a (mathematically real) *proper distance* interval $\Delta\sigma$ for this pair of events? (d) Would you describe this pair of events as timelike? Spacelike? Lightlike?

**45. Reversing an argument.**   In our physical derivation of the length contraction of a moving rod (Section 2.1), we assumed that the time dilation was given. In a similar manner, derive the time dilation for a moving clock, assuming that the length contraction is given.

**46. Length contraction—another approach.**   We could define the length of a moving rod as the product of its velocity by the time interval between the instant that one end point of the rod passes a fixed marker and the instant the other end point passes the same marker. Show that this definition also leads to the length contraction result of Eq. 2-12b. (*Hint:* Let the rod be at rest in frame $S'$ and let the marker be fixed at one position in frame $S$.)

**47. In a relativistic world, earth satellites are slow movers.**   To circle the earth in low orbit a satellite must have a speed of about 17,000 mi/h. Suppose that two such satellites orbit the earth in opposite directions. (a) What is their relative speed as they pass? Evaluate using the classical Galilean velocity transformation equation. (b) What fractional error was made because the (correct) relativistic transformation equation was not used?

**48. Double checking.**   (a) Derive Eq. 2-22a the way Eq. 2-22b was derived. (b) In Table 2-4 we arrived at the inverse velocity transformation relations (right-hand column) by changing the sign of $v$ and interchanging the primed and unprimed quantities in the equations in the left-hand column. Verify this procedure by deriving the equation for $u'_y$ directly, using the same procedure that we used in deriving the equation for $u_y$ (that is, Eq. 2-22b).

**49. A moving particle.**   A particle moves along the $x'$ axis of frame $S'$ with a speed of $0.40c$. Frame $S'$ moves with a speed of $0.60c$ with respect to frame $S$. What is the measured speed of the particle in frame $S$?

**50. $S$ and $S'$ watch a moving particle.**   Frame $S'$ moves relative to frame $S$ at $0.60c$ in the direction of increasing $x$. In frame $S'$ a particle is measured to have a velocity of $0.40c$ in the direction of increasing $x'$. (a) What is the velocity of the particle with respect to frame $S$? (b) What would be the velocity of the particle with respect to $S$ if it moved (at $0.40c$) in the direction of *decreasing $x'$* in the $S'$ frame? In each case, compare your answers with the predictions of the classical velocity transformation equation.

**51. Two fast particles rush toward each other.**   One cosmic-ray particle approaches the earth along its axis with a velocity of $0.80c$ toward the North Pole and another, with a velocity of $0.60c$, toward the South Pole. What is the relative speed of approach of one particle with respect to the other? (*Hint:* It is useful to consider the earth and one of the particles as the two inertial reference frames.)

**52. Interesting information but not very helpful.** The meteorite watch officer on a spaceship reports that two fast micrometeorites are approaching the ship on parallel tracks, one at a speed of $0.90c$ and the other at $0.70c$. What is the speed of either of them with respect to the other?

**53. Faster than a speeding bullet!** A spaceship whose rest length is 300.0 m has a speed of $0.80c$ with respect to a certain reference frame. A micrometeorite, also with a speed of $0.80c$ in this frame, passes the spaceship on an antiparallel track. How long does it take this object to pass the spaceship?

**54. The expanding universe (I).** Galaxy A is reported to be receding from us with a speed of $0.35c$. Galaxy B, located in precisely the opposite direction, is also found to be receding from us at this same speed. What recessional speed would an observer on Galaxy A find (*a*) for our galaxy? (*b*) For Galaxy B?

**55. The expanding universe (II).** It is concluded from measurements of the red shift of the emitted light that quasar $Q_1$ is moving away from us at a speed of $0.80c$. Quasar $Q_2$, which lies in the same direction in space but is closer to us, is moving away from us at speed $0.40c$. What velocity for $Q_2$ would be measured by an observer on $Q_1$?

**56. Fast, but not as fast as you might think.** Starfleet spacecruisers *Lorentz* and *Minkowski* are proceeding outward from Lunar Base on a straight line with *Lorentz* leading. *Lorentz* has a speed of $0.60c$ with respect to *Minkowski*, which, in turn, has a speed of $0.60c$ with respect to Lunar Base. What is the speed of *Lorentz* with respect to Lunar Base?

**57. The unreachable goal!** A spaceship, at rest in a certain reference frame $S$, is given a speed increment of $0.50c$. It is then given a further $0.50c$ increment in this new frame, and this process is continued until its speed

with respect to its original frame $S$ exceeds $0.999c$. How many increments does it require?

**58. Directions change too.** A particle moves with speed $u$ at an angle $\theta$ with respect to the $x$ axis in frame $S$. Frame $S'$ moves along this axis with speed $v$. What speed $u'$ and angle $\theta'$ will the particle appear to have to an observer in $S'$?

**59. Watching the decay of a moving nucleus.** A radioactive nucleus moving with a uniform velocity of $0.050c$ along the $x$ axis of a reference frame ($S$) fixed with respect to the laboratory. It decays by emitting an electron whose speed, measured in a reference frame ($S'$) moving with the nucleus, is $0.800c$. Consider first the cases in which the emitted electron travels (*a*) along the common $x$-$x'$ axis and (*b*) along the $y'$ axis and find, for each case, its velocity (magnitude and direction) as measured in frame $S$. (*c*) Suppose, however, that the emitted electron, viewed now from frame $S$, travels along the $y$ axis of that frame with a speed of $0.800c$. What is its velocity (magnitude and direction) as measured in frame $S'$?

**60. A philosophical difficulty.** Suppose that event $A$ *causes* event $B$, the effect now being propagated from $A$ to $B$ with a speed greater than $c$. Show, using the relativistic velocity transformation equation, that there exists an inertial frame $S'$, which moves relative to $S$ with a velocity less than $c$, in which the order of these events would be reversed. Hence, if concepts of cause and effect are to be preserved, it is impossible to send signals with a speed greater than that of light.

**★61. A surprising result.** In Fig. 2-17, $A$ and $B$ are trains on perpendicular tracks, shown radiating from station $S$. The velocities are in the station frame ($S$ frame). (*a*) Find $\mathbf{v}_{AB}$, the velocity of train $B$ with respect to train $A$. (*b*) Find $\mathbf{v}_{BA}$, the velocity of train $A$ with respect to train $B$. (*c*) Comment on the fact that these two relative velocities do not point in opposite directions [12].

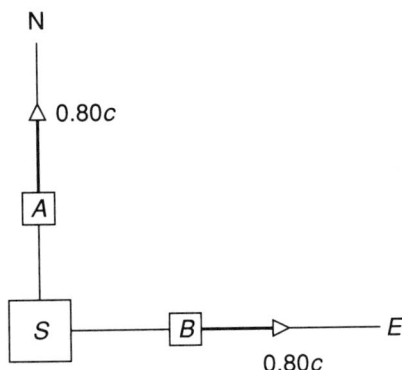

**FIGURE 2-17.**    Problem 61.

62. **A neat formulation.**   A particle has a speed $u$ in frame $S$ and a corresponding speed $u'$ in frame $S'$, where

$$u^2 = u_x^2 + u_y^2$$

and

$$u'^2 = u_x'^2 + u_y'^2.$$

(a) Verify by direct substitution and by use of the appropriate velocity transformation equations from Table 2-4 that the following relationship holds:

$$(c^2 - u^2)(c^2 + u_x'v)^2 = c^2(c^2 - u'^2)$$
$$\times (c^2 - v_2).$$

(b) Show that this formulation contains within itself the result that if $u' < c$ and also $v < c$ then $u$ must be less than $c$. (c) Show also that the equation contains the result that if $u' = c$ or if $v = c$, then $u$ must also be equal to $c$. See [13].

63. **How to put a long pole into a short garage.**   Suppose that a pole vaulter, holding a 16-ft pole parallel to his direction of motion, runs toward an open garage that is 8 ft deep. The far end of the garage is a massive concrete barrier. (a) At what speed must the pole vaulter run if, at the instant the front end of the pole touches the barrier, the rear end is within the garage entrance, so that the entire pole is contained within the garage? (b) In the preceding we assumed a reference frame in which the garage was at rest. Consider,

however, a reference frame fixed with respect to the pole. In this frame the pole has its rest length ($= 16$ ft), but the garage, which now rushes toward the runner, is contracted (to 4 ft!) by the same Lorentz factor that operated on the pole in (a). How can a 16-ft pole fit into a 4-ft garage? If it can't, there is a violation of the principle of relativity, a serious matter indeed. (*Hint:* No body is truly rigid. When the front of the pole hits the barrier, the rear end of the pole keeps on going, at unchanged speed, until it "gets the word" by means of a compression wave sent down the pole. The question is, is there enough time for the rear end of the pole to get inside the garage before the front end of the pole reaches it? Can you show that the answer is "yes"? See [14].)

64. **A very large Doppler shift.**   A spaceship, moving away from the earth at a speed of $0.90c$, reports back by transmitting on a frequency (measured in the spaceship frame) of 100 MHz. To what frequency must earth receivers be tuned to receive these signals?

65. **The Doppler shift in terms of wavelengths.**   Show that the Doppler shift formulas (Eqs. 2-30) can be written in the forms

$$\lambda = \lambda_0(1 - \beta + \tfrac{1}{2}\beta^2 - \cdots)$$

(approaching)

and

$$\lambda = \lambda_0(1 + \beta + \tfrac{1}{2}\beta^2 + \cdots)$$

(separating),

in which $\lambda_0$ is the proper wavelength, that is, the wavelength measured by an observer for whom the source is at rest. These formulations are especially useful for $\beta \ll 1$. Compare Eqs. 2-31.

66. **A red-shifted quasar.**   Observations on the light from a certain quasar show a red shift of a spectral line of laboratory wavelength 500 nm to a wavelength of 130 nm. What is the quasar's speed of recession from us, according to a Doppler-effect interpretation?

67. **Doppler shifts and the rotating sun.**   Because of the rotation of the sun, points on its

surface at its equator have a speed of 1.85 km/s with respect to its center. Consider groups of atoms on opposite edges of the sun's equator as seen from the earth, emitting light of proper wavelength 546 nm. What wavelength difference is observed on the earth for the light from these two groups of atoms? (*Hint:* Use the formulas displayed in Problem 65.)

68. **A Doppler shift revealed as a color change.** A spaceship is receding from the earth at a speed of 0.20c. A light on the rear of the ship appears blue ($\lambda = 450$ nm) to passengers on the ship. What color would it appear to an observer on earth?

69. **The exact and the approximate Doppler formulas compared.** (*a*) Calculate the Doppler wavelength shifts $\lambda - \lambda_0$ expected for the sodium $D_1$ line ($\lambda_0 = 589.6$ nm) for source and observer approaching each other at relative speeds of 0.050c, 0.40c, and 0.80c. (*b*) Calculate the same quantities using the formula developed in Problem 65, discarding terms of order $\beta^2$ or higher. Compare the two sets of results.

70. **Quasar, quasar, burning bright. . . .** In the spectrum of quasar 3C9, some of the familiar hydrogen lines appear but they are shifted so far forward toward the red that their wavelengths are observed to be three times as large as that observed in the light from hydrogen atoms at rest in the laboratory. (*a*) Show that the classical Doppler equation gives a velocity of recession greater than c. (*b*) Assuming that the relative motion of 3C9 and the earth is entirely one of recession, find the recession speed predicted by the relativistic Doppler equation.

71. **The Ives-Stillwell experiment.** Neutral hydrogen atoms are moving along the axis of an evacuated tube with a speed of $2.0 \times 10^6$ m/s. A spectrometer is arranged to receive light emitted by these atoms in the direction of their forward motion. This light, if emitted from resting hydrogen atoms, would have a measured (proper) wavelength of 486.133 nm. (*a*) Calculate the expected wavelength for light emitted from the forward-moving (approaching) atoms, using the exact relativistic formula (see Eqs. 2-30). (*b*) By use of a mirror this same spectrometer can also measure the wavelength of light emitted by these moving atoms in the direction opposite to their motion. What wavelength is expected under this arrangement, in which the light source and the observer are—effectively—separating? (*c*) Calculate the difference between the average of the two wavelengths found in (*a*) and (*b*) and the unshifted (proper) wavelength. Show, by analyzing the formulas displayed in Problem 65, that this difference measures the $\beta^2$ term in these formulas. By this technique, Ives and Stillwell (see *Physics*, Part II, Sec. 42-5) were able to distinguish between the predictions of the classical and the relativistic Doppler formulas.

72. **The transverse Doppler effect.** Show that the transverse Doppler shift formula (Eq. 2-32) can be written in the form

$$\lambda = \lambda_0(1 + \tfrac{1}{2}\beta^2 + \tfrac{3}{8}\beta^4 + \cdots),$$

in which $\lambda_0$ is the proper wavelength, that is, the wavelength that would be measured by an observer for whom the source is at rest. This formulation is especially useful for $\beta \ll 1$. Compare Problem 65.

73. **A case of purely transverse motion.** Give the Doppler wavelength shift $\lambda - \lambda_0$, if any, for the sodium $D_2$ line (589.00 nm) emitted from a source moving in a circle with constant speed (= 0.10c) as measured by an observer fixed at the center of the circle.

74. **A case of (not quite) transverse motion.** Calculate the wavelength shift

$$\Delta\lambda = \lambda - \lambda_0$$

for $\lambda_0 = 589.00$ nm, $\beta = 0.10$ and (*a*) $\theta = 90°$ (*b*) $88°$ (*c*) $85°$. (*d*) Why does such a small departure from purely transverse motion generate such a large change in the measured Doppler shift? What lessons are there

here for the experimenter who wishes to measure the transverse Doppler effect?

**75. The Doppler effects for sound and light compared.**   In the case of wave propagation in a medium (sound in air, say), the Doppler shifts for the source moving through the medium and for the observer moving through the medium are different, even though the relative speeds of the source with respect to the observer may be the same. For light in free space, however, the two situations are completely equivalent; the same Doppler shift results. Show that if we take the geometric mean of the two former results, we get exactly the relativistic Doppler shift of Eq. 2-27. (See *Physics*, Secs. 20-7 and 42-5; recall that the geometric mean of two quantities, $a$ and $b$, is $\sqrt{ab}$.)

**76. A moving radar transmitter and a moving clock.**   A radar transmitter $T$ is fixed to a reference frame $S'$ that is moving to the right with speed $v$ relative to reference frame $S$ (see Fig. 2-18). A mechanical timer (essentially a clock) in frame $S'$, having a period $\tau_0$ (measured in $S'$) causes transmitter $T$ to emit radar pulses, which travel at the speed of light and are received by $R$, a receiver fixed in frame $S$. (*a*) What would be the period $\tau$ of the timer relative to observer $A$, who is fixed in frame $S$? (*b*) Show that the receiver $R$ would observe the time interval between pulses arriving from $T$, not as $\tau$ or as $\tau_0$, but as

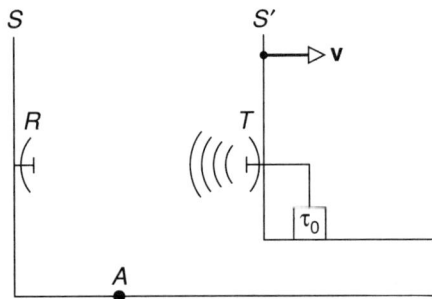

$$\tau_R = \tau_0 \sqrt{\frac{c + v}{c - v}}.$$

(*c*) Explain why the observer at $R$ measures a different period for the transmitter than does observer $A$, who is in the same reference frame. (*Hint:* A clock and a radar pulse are not the same.)

**77. The classical aberration formula.**   In Example 11 the aberration of light from a star that is directly overhead is shown to be given (for $\beta \ll 1$) by

$$\tan \theta' = \frac{1}{\beta}.$$

The aberration angle $\alpha$ is shown in Fig. 2-11 to be related to $\theta'$ by

$$\alpha = 270° - \theta'.$$

Show, by combining these relations, that

$$\tan \alpha = \beta$$

results. This (see Eq. 1-11) is the prediction of classical theory for the aberration of starlight.

**78. A moving rod and a moving laser beam.**   (*a*) A rod makes an angle of 30° with the $x'$ axis of frame $S'$, which is moving with speed $0.80c$ with respect to frame $S$. What angle does the rod make with the $x$ axis of frame $S$? (*b*) A laser beam, generated by a laser gun fixed in frame $S'$, also makes an angle of 30° with the $x'$ axis. What angle does it make with the $x$ axis, as determined by an observer in frame $S$? Why are these angles so different?

**79. A particle and a light pulse.**   A particle has a speed $u'$ in the $S'$ frame, its track making an angle $\theta'$ with the $x'$ axis. The particle is viewed by an observer in frame $S$, the two frames having a relative speed parameter $\beta$. (*a*) Show that the angle $\theta$ made by the track of the particle with the $x$ axis is given by

$$\tan \theta = \frac{u' \sin \theta'}{\gamma(u' \cos \theta' + \beta c)}.$$

(*b*) Show that this equation reduces to the standard aberration formula (Eq. 2-29*a*) if

**FIGURE 2-18.**   Problem 76.

the "particle" is, in fact, a light pulse, so that $u' = c$.

80. **The aberration of light—a different formulation.** Show that, by combining Eqs. 2-25 and 2-27 (rather than 2-25 and 2-26), the aberration formula can be written in the form

$$\cos \theta = \frac{\cos \theta' + \beta}{1 + \beta \cos \theta'}.$$

Test that this formula is equivalent to Eq. 2-29a by finding the value of $\theta$ corresponding to $\theta' = 30°$ and $\beta = 0.80$, using each formula.

81. **A moving nucleus emits a gamma ray.** A radioactive nucleus moves with a uniform velocity of $0.050c$ in the laboratory frame. It decays by emitting a gamma ray, which we may view as a pulse of electromagnetic radiation. What are the magnitude and direction of the velocity of this pulse as observed in the laboratory frame? Assume that the pulse is emitted (a) parallel to the direction of the motion of the nucleus, as judged by an observer on the nucleus; (b) at 45° to this direction; (c) at right angles to this direction.

82. **The headlight effect.** A source of light, at rest in the $S'$ frame, emits radiation uniformly in all directions. (a) Show that the fraction of light emitted into a cone of half-angle $\theta'$ is given by

$$f = 0.50(1 - \cos \theta').$$

Calculate $f$ for $\theta' = 30°$. (b) The source is viewed from frame $S$, the relative velocity of the two frames being $0.80c$. Find the value of $\theta$ (in frame $S$) to which this value of $f$ corresponds, using the appropriate aberration formula. Repeat the calculation for $\beta = 0.90$ and for $\beta = 0.990$. Can you see why this aberration phenomenon is often referred to as the "headlight effect"?

83. **The headlight effect—a high-speed limit.** A source of light, at rest in the $S'$ frame, emits uniformly in all directions. The source is viewed from frame S, the relative speed

parameter relating the two frames being $\beta$. (a) Show that at high speeds (that is, as $\beta \rightarrow 1$), the forward- pointing cone into which the source emits half of its radiation has a half-angle $\theta_{0.5}$ given closely, in radian measure, by

$$\theta_{0.5} = \sqrt{2(1 - \beta)}.$$

(b) What value of $\theta_{0.5}$ is predicted for the gamma radiation emitted by a beam of energetic neutral pions, for which $\beta = 0.993$? (c) At what speed would a light source have to move toward an observer to have half of its radiation concentrated into a narrow forward cone of half-angle 5.0°?

84C. **Using your calculator—Lorentz transformations for an event pair.** Write a program for your handheld, programmable calculator to handle Lorentz transformations for an event pair. Accept as inputs: (1) the speed parameter $\beta$, (2) the $\Delta x$ (or the $\Delta x'$) coordinate difference, and (3) the $\Delta t$ (or the $\Delta t'$) coordinate difference. Display as outputs, in succession: (1) the Lorentz factor $\gamma$, (2) the $\Delta x'$ (or the $\Delta x$) coordinate difference, (3) the $\Delta t'$ (or the $\Delta t$) coordinate difference, (4) the spacetime interval, and (5) a signal as to whether the event pair is spacelike or timelike. (*Hint:* See Table 2-3. Also, if your calculator does not display letters you can manage (5) above by displaying a string of 5's for "spacelike" and a string of 7's for "timelike".)

85C. **Using your calculator.** Test the program that you have written in Problem 84C in the following ways. (a) Two events are simultaneous in frame $S$. Show, by trial, that they cannot be simultaneous in frame $S'$ no matter what values you assign to $\beta$ (except zero) or to $\Delta x$ (except zero). Show also, by trial, that this event pair will be spacelike in all frames. (b) Two events occur at the same position in frame $S$. Show, by trial, that they cannot occur at the same position in frame $S'$ no matter what values you assign to $\beta$ (except zero) or to $\Delta t$ (except zero). Show, also by trial, that this event pair will be timelike in all frames. (c) A rod is at rest in frame $S'$. You (in frame $S$) measure its length to be 5.00 m by making simultaneous measurements of its endpoints. What is its rest length if $\beta = 0.600$? Is your result consistent with the length contraction phenomenon? (d) A clock is at rest in frame $S$. You (also in frame $S$) measure a time interval of 2.50 $\mu$s between two events. What interval would $S'$ observe between these same events if $\beta = 0.600$? Is your result consistent with the time dilation phenomenon? (e) Put $\Delta x = 5.00 \times 10^8$ m and $\Delta t = 4.00$ s. What is the magnitude and the nature of the spacetime interval associated with this event pair? Show, by experimenting with various values of $\beta$ (including zero and negative values) that the interval is the same in all inertial frames. Change the character of the interval by changing $\Delta x$ and by changing $\Delta t$.

**86C. Using your calculator—the Doppler effect.** Use your handheld, programmable calculator to write a program that will do the following: Accept as inputs (1) the proper wavelength $\lambda_0$ of the source *or* (2) the corresponding proper frequency $\nu_0$, and also (3) the speed parameter $\beta$; if the source and the observer are approaching put $\beta > 0$ but if they are separating put $\beta < 0$. Display as successive outputs (1) the wavelength $\lambda$, (2) the proper wavelength $\lambda_0$, (3) the wavelength shift $(\lambda - \lambda_0)$, (4) the frequency $\nu$, (5) the proper frequency $\nu_0$, and (6) the frequency shift $(\nu - \nu_0)$. See Eqs. 2-30.

**87C. Using your calculator.** Check out the program that you have written in Problem 86C as follows: (a) A galaxy is receding from us at $0.30c$ where $c$ is the speed of light. What is the expected shift in wavelength for the sodium spectrum line in the light from this galaxy? The laboratory wavelength of this line is 589 nm. (b) An automobile is receding from a Doppler radar speed detector at 100 mi/hr ($= 44.7$ m/s). What will be the frequency shift of the reflected radar beam? The proper wavelength of the radar radiation is 3.00 cm.

# REFERENCES

1. ROBERT RESNICK, *Introduction to Special Relativity* (Wiley, New York, 1968).
2. DAVID H. FRISCH AND JAMES H. SMITH, "Measurement of Relativistic Time Dilation Using $\mu$-Mesons," *Am. J. Phys.*, **31**, 342 (1963). See also the related film, "Time Dilation—An Experiment with $\mu$-Mesons," Educational Services, Inc., Watertown, Mass.
3. V. T. WEISSKOPF, "The Visual Appearance of Rapidly Moving Objects," *Phys. Today* (September 1960).
4. N. C. McGILL, "The Apparent Shape of Rapidly Moving Objects in Special Relativity," *Contemp. Phys.* (January 1968).
5. G. D. SCOTT AND H. J. VAN DRIEL, "Geometric Appearances at Relativistic Speeds," *Am J. Phys.*, **38**, 971 (1970).
6. MILTON A. ROTHMAN, "Things That Go Faster Than Light," *Scientific American* (July 1960).
7. GERALD FEINBERG, "Particles That Go Faster Than Light," *Scientific American* (February 1970).
8. HIRSCH I. MANDELBERG AND LOUIS WITTEN, "Experimental Verification of the Relativistic Doppler Effect," *J. Optical Soc. Am.*, **52**, 529 (1962).
9. WALTER KUNDIG, "Measurement of the Transverse Doppler Effect in an Accelerated System," *Phys. Rev.*, **129**, 2371 (1963).
10. H. BONDI, "The Teaching of Special Relativity," *Phys. Educ.*, **1**, 223 (1966).
11. J. C. HAFELE AND RICHARD E. KEATING, "Around-the-World Atomic Clocks: Predicted Relativistic Time Gains," *Science*, **177**, 166 (1972), and J. C. Hafele and Richard E. Keating, "Around-

the-World Atomic Clocks: Observed Relativistic Time Gains,'' *Science,* **177,** 168 (1972).

12. Suggested by Professor William Doyle of Dartmouth College.

13. See Wolfgang Rindler, *Essential Relativity* (Van Nostrand Reinhold, New York, 1969), sec. 36.

14. Wolfgang Rindler, ''Length Contraction Paradox,'' *Am. J. Phys.,* **29,** 365 (1961).

15. N. David Mermin, ''Relativistic Addition of Velocities Directly from the Constancy of the Velocity of Light,'' *Am. J. Phys.,* **51,** 1130 (1983).

16. Harry Woolf, Ed. *Some Strangeness in the Proportion: A Centennial Symposium to Celebrate the Achievements of Albert Einstein* (Addison-Wesley, Reading, Mass., 1980). See especially Wolfgang K. H. Panofsky, ''Special Relativity in Engineering,'' and Edward M. Purcell, ''Comments on 'Special Relativity Theory in Engineering.' ''

## CHAPTER 3

# Relativistic Dynamics

*From this equation it directly follows that:—If a body gives off the energy* L *in the form of radiation, its mass diminishes by* L/c². *The fact that the energy withdrawn from the body becomes energy of radiation evidently makes no difference, so that we are led to the more general conclusion that . . . The mass of a body is a measure of its energy content. . . .*

*Albert Einstein (1905)*

## 3-1 Mechanics and Relativity

In Chapter 1 we saw that experiment forced us to the conclusion that the Galilean transformations had to be replaced and the basic laws of mechanics, which were consistent with those transformations, needed to be modified. In Chapter 2 we obtained the new transformation equations, the Lorentz transformations, and examined their implications for kinematic phenomena. Now we must consider dynamic phenomena and find how to modify the laws of classical mechanics so that the new mechanics is consistent with relativity.

Basically, classical Newtonian mechanics is inconsistent with relativity because its laws are invariant under a Galilean transformation and *not* under a Lorentz transformation. This formal result is plausible, as well, from other considerations. For example, in Newtonian mechanics a force can accelerate a particle to indefinite speeds, whereas in relativity the limiting speed is $c$. We need a new law of motion that is consistent with relativity. When we obtain such a law of motion, we must also ensure that it reduces to the Newtonian form as $\beta \; (= v/c) \to 0$, since, in the domain where $\beta \ll 1$, Newton's laws are consistent with experiment. Thus, the relativistic law of motion will be a generalization of the classical one.

We shall proceed by studying collisions. The laws of conservation of momentum and energy are valid classically during such interactions. If we require that these conservation laws also be valid relativistically (that is, invariant under a Lorentz transformation) and hence that they be general laws of physics, we must modify them from the classical form in such a way that they also reduce to the classical form as $\beta \to 0$. In this way, we shall obtain the relativistic law of motion.

## 3-2 The Need to Redefine Momentum

The first thing we wish to show is that if we want to find a quantity such as momentum (for which there is a conservation law in classical physics) that is also subject to a conservation law in relativity, we cannot use the same expression for momentum as the classical one. We must, instead, redefine momentum in order that a law of conservation of momentum in collisions be invariant under a Lorentz transformation. Ultimately this leads to a more general view of the concept of mass as well.

Let us begin by considering a head-on collision between two identical particles. We choose the simplest case, namely, one in which the particles stick together after impact—a so-called totally inelastic collision. We shall first analyze the collision from the point of view of classical mechanics (see *Physics*, Part I, Sect. 10-4).

In Fig. 3-1a we view the collision from a reference frame $S$ chosen so as to present the collision as symmetrically as possible. Initially, we see that the particles, which have the same mass $m$, are approaching each other with the same speed $u$. After the collision the particles form one larger particle whose mass is $2m$ (because of the conservation of mass) and whose speed in the $S$ frame is zero (because of the conservation of momentum). The total system momentum $p$, as measured in frame $S$, is zero both before and after the collision.

Let us now view this same collision from a frame $S'$, moving to the left as viewed from $S$ with a speed $v$ that we deliberately choose to be equal to $u$; see Fig. 3-1b. With this choice for the separation speed of the two frames, *one* of the colliding particles is initially at rest in $S'$ and the resulting compound particle moves to the right with speed $u$. The other colliding particle has a speed indicated by $u'$ in Fig. 3-1b. From the

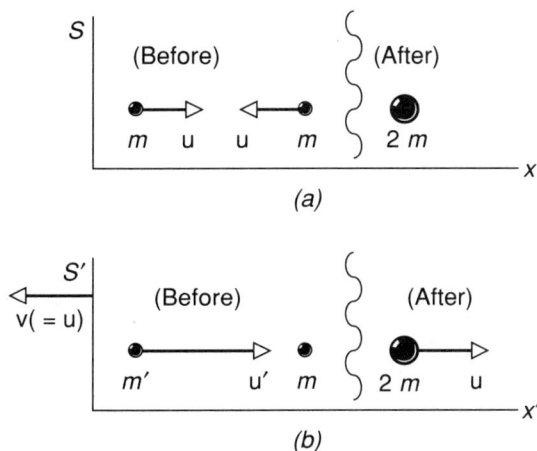

**FIGURE 3-1.** A totally inelastic collision between two identical particles is viewed from two reference frames. (a) Frame $S$, in which the center of mass of the system is stationary. (b) Frame $S'$, which moves to the left with respect to frame $S$ with speed $v(= u)$.

Galilean velocity transformation law, however, we see at once that $u' = 2u$. Note that again the mass is conserved (at the value $2m$) and so is the total system momentum (now at the value $p' = 2mu$) in frame $S'$.

Now let us examine this same collision from the relativistic point of view. We shall carry out the analysis just as before with the exception that we shall use the relativistic velocity transformation law in place of the classical Galilean transformation law we used earlier. In frame $S$ momentum is still conserved, simply because of the symmetry of the collision as viewed from this frame. In frame $S'$, however, the speed shown as $u'$ in Fig. 3-1b is no longer equal to $2u$, but is given by (see Eq. 2-21)

$$u' = \frac{u + v}{1 + uv/c^2} = \frac{2u}{1 + u^2/c^2}.$$

After the collision the speed of the compound particle will again be $u$. We see that, although mass is again conserved at the value $2m$, the total momentum is *not* conserved in frame $S'$. Before the collision it is $p' = mu' = 2mu/(1 + u^2/c^2)$, whereas after the collision it is simply $p' = 2mu$.

Our conclusion is that, if we compute momentum according to the classical formulas $\mathbf{p} = m\mathbf{u}$ and $\mathbf{p}' = m\mathbf{u}'$, then when momentum is conserved in a collision in one frame, it is not conserved in the other frame. This result contradicts the basic postulate of special relativity that the laws of physics are the same in all inertial reference frames. If the conservation of momentum in collisions is to be a law of physics, then the classical definition of momentum cannot be correct in general.

In the next section, we shall show that it is possible to preserve the *form* of the classical definition of the momentum of a particle, $\mathbf{p} = m\mathbf{u}$, where $\mathbf{p}$ is the momentum, $m$ the mass, and $\mathbf{u}$ the velocity of a particle, and also to preserve the classical law of the conservation of momentum of a system of interacting particles, providing that we modify the classical concept of mass. We need to let the mass of a particle be a function of its speed $u$; that is, $m = m_0/\sqrt{1 - u^2/c^2}$, where $m_0$ the classical mass and $m$ is the relativistic mass of the particle. Clearly, as $u/c$ tends to zero, $m$ tends to $m_0$. The relativistic momentum then becomes $\mathbf{p} = m\mathbf{u} = m_0\mathbf{u}/\sqrt{1 - \beta^2}$ and reduces to the classical expression $\mathbf{p} = m_0\mathbf{u}$ as $\beta \rightarrow 0$. Let us now deduce these results.

## 3-3 Relativistic Momentum

In our analysis above, we considered the mass of a particle to be independent of its motion. However, we have already learned that the measured length of a rod and the measured rate of a clock are affected by the motion of the rod or the clock relative to an observer. Would it be so surprising, then, to find that the measured mass of a particle also depends on its state of motion?

Let us then assume that the mass of a particle is *not* a constant but is a function of the speed of the particle. Then, in the totally inelastic collision we have been considering, we should use the symbol $m_0$ for the mass of a particle at rest, $m$ for the mass of the same particle if its speed is $u$, and $m'$ for its mass if its speed is $u'$.

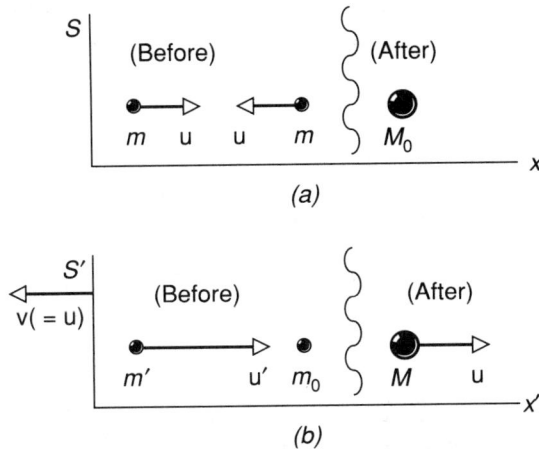

**FIGURE 3-2.** The same collision as in Fig. 3-1, with the masses and speeds relabeled with symbols appropriate for a relativistic analysis.

For the masses of the compound particle we can use $M_0$ and $M$. Figure 3-2 shows the collision of Fig. 3-1 with the masses and the speeds relabeled in this way.

Because we assume that (relativistic) mass is conserved in the $S'$ frame of Fig. 3-2b, we can write

$$M = m' + m_0, \tag{3-1}$$

and, for the conservation of (relativistic) momentum,

$$Mu = m'u'. \tag{3-2}$$

Now $u'$ is related to $u$ by the relativistic law for the addition of velocities, or (see Eq. 2-21)

$$u' = \frac{u + v}{1 + uv/c^2}.$$

Recalling that $v = u$ (see Fig. 3-2), we can rewrite this as

$$u' = \frac{2u}{1 + u^2/c^2}. \tag{3-3}$$

The above three numbered equations involve five variables, $m_0$, $m'$, $M$, $u'$, and $u$. If we could eliminate $M$ and $u$ from these equations, we would be left with a relationship involving only $m'$, $m_0$, and $u'$. If we look at the left-hand particle in Fig. 3-2b, we see that these are precisely the variables we need if we are to relate the relativistic mass ($m'$) of this particle to its rest mass ($m_0$) and to its speed ($u'$).

We start by eliminating $M$ between Eqs. 3-1 and 3-2, obtaining

$$u = u' \left( \frac{m'}{m' + m_0} \right).$$

Combining this with Eq. 3-3 leads, after some rearrangement (see Problem 18), to

$$m' = \frac{m_0}{\sqrt{1 - (u')^2/c^2}},$$

which is the relationship we seek. We can extend our conclusion beyond the confines of the specific problem we have been analyzing and write as a general result, true for *any* particle of rest mass $m_0$ and speed *u:*

$$m = \frac{m_0}{\sqrt{1 - u^2/c^2}}, \qquad (3\text{-}4a)$$

which tells us how the relativistic mass $m$ of a body varies with its speed $u$. If we represent $u/c$ by $\beta$, we can express this relationship as

$$m = \frac{m_0}{\sqrt{1 - \beta^2}} \qquad (3\text{-}4b)$$

or, equivalently, as

$$m = m_0\gamma, \qquad (3\text{-}4c)$$

in which $\gamma$ is the familiar Lorentz factor. Recall that this factor enters in a similar simple and direct way into the relativistic formulas for length contraction (see Eq. 2-12b) and time dilation (see Eq. 2-14b).

Inspection of Eqs. 3-4 shows us that (1) the relativistic mass $m$ is always greater than the rest mass $m_0$, and (2) as $\beta \to 0$ (or, equivalently, as $\gamma \to 1$), then $m \to m_0$, as we expect in this Newtonian low-speed limit.

Hence, if we want to preserve the *form* of the classical momentum conservation law while requiring that the law be relativistically invariant, we must define the mass of a moving body by Eqs. 3-4. That is, momentum still has the form $m\mathbf{u}$, but mass is defined as $m = m_0/\sqrt{1 - u^2/c^2}$. Note that $u$ is the speed of the body relative to $S$, which we can regard as the laboratory frame, and that $u$ has no necessary connection with changing reference frames. By accepting Eqs. 3-4 as our definition of the mass of a moving body, we implicitly assume that the mass of a body does not depend on its acceleration relative to the reference frame, although it does depend on its speed. Mass remains a scalar quantity in the sense that its value is independent of the *direction* of the velocity of the body. The rest mass $m_0$ is often called the *proper mass,* for it is the mass of the body measured, like proper length and proper time, in the inertial frame in which the body is at rest.

We have presented above a derivation of an expression for relativistic momentum that obviously centers around a very special case, a totally inelastic one-dimensional collision. Such a derivation enables us to make an educated guess as to what the general result may be. We have avoided rather involved general derivations that lead, in any case, to exactly the same results. When the general case is done, $u$ becomes the absolute value of the velocity of the particle; that is, $u^2 = u_x^2 + u_y^2 + u_z^2$.

Hence, to conclude, in order to make the conservation of momentum in collisions a law that is experimentally valid in all reference frames, we must define momentum not as $m_0\mathbf{u}$, but as

$$\mathbf{p} = \frac{m_0\mathbf{u}}{\sqrt{1 - u^2/c^2}}. \qquad (3\text{-}5)$$

The components of the momentum then are

$$p_x = \frac{m_0 u_x}{\sqrt{1 - u^2/c^2}}, \quad p_y = \frac{m_0 u_y}{\sqrt{1 - u^2/c^2}}, \quad p_z = \frac{m_0 u_z}{\sqrt{1 - u^2/c^2}}, \qquad (3\text{-}6)$$

which we write out explicitly to emphasize that the magnitude $u$ of the total velocity appears in the denominator of each component equation.

## EXAMPLE 1.

***Relativistic Mass and Rest Mass.*** For what value of $u/c\ (=\beta)$ will the relativistic mass of a particle exceed its rest mass by a given fraction $f$? From Eq. 3-4$b$, we have

$$f = \frac{m - m_0}{m_0} = \frac{m}{m_0} - 1 = \frac{1}{\sqrt{1 - \beta^2}} - 1,$$

which, solved for $\beta$, is

$$\beta = \frac{\sqrt{f(2 + f)}}{1 + f}.$$

The table below shows some computed values, which hold for all particles regardless of their rest mass.

| $f$ | $\beta$ |
|---|---|
| 0.001 (0.1%) | 0.045 |
| 0.01 | 0.14 |
| 0.1 | 0.42 |
| 1 (100%) | 0.87 |
| 10 | 0.996 |
| 100 | 0.99995 |
| 1000 | 0.9999995 |

## EXAMPLE 2.

***Relativistic Mass Is Conserved.*** In writing Eq. 3-1 we assumed that relativistic mass is conserved when the collision of Fig. 3-2 is viewed from the $S'$ frame. Show that relativistic mass is also conserved in the $S$ frame.

The conservation of mass in the $S$ frame requires that

$$M_0 = m + m = 2m,$$

in which the meanings of the symbols will be clear from Fig. 3-2. Now $M_0$ is related to $M$, and $m$ to $m_0$, by the relativistic mass relationship (Eq. 3-4$a$). Using that equation, we can recast the above expression as

$$M = \frac{2m_0}{1 - u^2/c^2}. \qquad (3\text{-}7)$$

To demonstrate that Eq. 3-7 is true, we turn to Eqs. 3-1 and 3-2 and eliminate $m'$ between them. The result is

$$M = \frac{m_0}{1 - u/u'}.$$

We now use Eq. 3-3 to eliminate $u'$ from this expression, obtaining

$$M = \frac{m_0}{1 - \frac{1}{2}(1 + u^2/c^2)} = \frac{2m_0}{1 - u^2/c^2},$$

which is precisely Eq. 3-7, the relationship we sought to prove.

The implications of our new relativistic definition of mass for the relationship between mass and energy are considered in later sections.

We can summarize the relativistic definition of momentum that we have introduced in this section by writing, for the $x$-coordinate,

$$p_x = m_0 \gamma u_x = [m_0 \gamma]\, dx/dt = m\, dx/dt$$

in which we have combined the Lorentz factor $\gamma$ with the rest mass $m_0$ to generate the relativistic mass $m$. In some more advanced treatments of relativity, however, the

Lorentz factor is combined with the time element $dt$ and the concept of relativistic mass is not introduced. Thus:

$$p_x = m_0 \gamma u_x = m_0[dx/(dt/\gamma)] = m_0 \, dx/d\tau,$$

in which $d\tau$ is an element of proper time; see Eq. 2-14$b$. This approach has the advantage that both $m_0$ (which in this treatment no longer requires a subscript) and $d\tau$ are invariant quantities. The choice between the two approaches reduces to a matter of taste. We believe that, in an introductory treatment, the pedagogic advantages of the first approach outweigh the formal advantages of the second (see Reference 6, Section 3-4). Note that in neither case is the relativistic definition of momentum in question.

## 3-4 The Relativistic Force Law and the Dynamics of a Single Particle

Newton's second law must now be generalized to

$$\mathbf{F} = \frac{d}{dt}(\mathbf{p}) = \frac{d}{dt}\left(\frac{m_0 \mathbf{u}}{\sqrt{1 - u^2/c^2}}\right) \tag{3-8}$$

in relativistic mechanics. When the law is written in this form, we can immediately deduce the law of the conservation of relativistic momentum from it; when $\mathbf{F}$ is zero, $\mathbf{p} = m_0\mathbf{u}/\sqrt{1 - u^2/c^2}$ must be a constant. In the absence of external forces, the momentum is conserved. Notice that this new form of the law, Eq. 3-8, is *not* equivalent to writing

$$\mathbf{F} = m\mathbf{a} = \left(\frac{m_0}{\sqrt{1 - u^2/c^2}}\right)\left(\frac{d\mathbf{u}}{dt}\right),$$

in which we simply multiply the acceleration by the relativistic mass.

We find that experiment agrees with Eq. 3-8. When, for example, we investigate the motion of high-speed charged particles, it is found that the equation correctly describing the motion is

$$q(\mathbf{E} + \mathbf{u} \times \mathbf{B}) = \frac{d}{dt}\left(\frac{m_0 \mathbf{u}}{\sqrt{1 - u^2/c^2}}\right), \tag{3-9}$$

which agrees with Eq. 3-8. Here, $q(\mathbf{E} + \mathbf{u} \times \mathbf{B})$ is the Lorentz electromagnetic force, in which $\mathbf{E}$ is the electric field, $\mathbf{B}$ is the magnetic field, and $\mathbf{u}$ is the particle velocity, all measured in the same reference frame, and $q$ and $m_0$ are constants that describe the electrical (charge) and inertial (rest mass) properties of the particle, respectively (see *Physics*, Part II, Sec. 33-2). Notice that the form of the Lorentz force law of classical electromagnetism remains valid relativistically, as we should expect from the discussion of Chapter 1.

Later we shall turn to the question of how forces transform from one Lorentz frame to another. For the moment, however, we confine ourselves to one reference frame (the laboratory frame) and develop other concepts in mechanics, such as work and energy, which follow from the relativistic expression for force (Eq. 3-8). We shall confine ourselves to the motion of a single particle. In succeeding sections we shall consider many-particle systems and conservation laws.

In Newtonian mechanics we define the kinetic energy, $K$, of a particle to be equal to the work done by an external force in increasing the speed of the particle from zero to some value $u$ (see *Physics*, Part I, Sec. 7-5). That is,

$$K = \int_{u=0}^{u=u} \mathbf{F} \cdot d\mathbf{l},$$

where $\mathbf{F} \cdot d\mathbf{l}$ is the work done by the force $\mathbf{F}$ in displacing the particle through $d\mathbf{l}$. For simplicity, we can limit the motion to one dimension—say, $x$—the three-dimensional case being an easy extension. Then, classically,

$$K = \int_{u=0}^{u=u} F\, dx = \int m_0 \left(\frac{du}{dt}\right) dx = \int m_0\, du\, \frac{dx}{dt} = m_0 \int_0^u u\, du = \tfrac{1}{2}m_0 u^2.$$

Here we write the particle mass as $m_0$ to emphasize that, in Newtonian mechanics, we do not regard the mass as varying with the speed, and we take the force to be $m_0 a = m_0(du/dt)$.

In relativistic mechanics, it proves useful to use a corresponding definition for kinetic energy in which, however, we use the relativistic equation of motion, Eq. 3-8, rather than the Newtonian one. Then, relativistically,

$$K = \int_{u=0}^{u=u} F\, dx = \int \frac{d}{dt}(mu)\, dx = \int d(mu)\, \frac{dx}{dt}$$

$$= \int (m\, du + u\, dm)u = \int_{u=0}^{u=u} (mu\, du + u^2\, dm), \tag{3-10}$$

in which both $m$ and $u$ are variables. These quantities are related, furthermore, by Eq. 3-4a, $m = m_0/\sqrt{1 - u^2/c^2}$, which we can rewrite as

$$m^2 c^2 - m^2 u^2 = m_0^2 c^2.$$

Taking differentials in this equation yields

$$2mc^2\, dm - m^2 2u\, du - u^2 2m\, dm = 0,$$

which, upon division by $2m$, can also be written as

$$mu\, du + u^2\, dm = c^2\, dm.$$

The left side of this equation is exactly the integrand of Eq. 3-10. Hence, we can write the relativistic expression for the kinetic energy of a particle as

$$K = \int_{u=0}^{u=u} c^2\, dm = c^2 \int_{m=m_0}^{m=m} dm = mc^2 - m_0 c^2$$

$$= (m - m_0)c^2. \tag{3-11a}$$

By using Eq. 3-4, we obtain equivalently

$$K = m_0 c^2 \left(\frac{1}{\sqrt{1 - u^2/c^2}} - 1\right) \tag{3-11b}$$

or

$$K = m_0 c^2 (\gamma - 1). \tag{3-11c}$$

Also, if we take $mc^2 = E$, where $E$ is called the *total energy* of the particle-a name whose aptness will become clear later—we can express Eq. 3-11a compactly as

$$E = m_0 c^2 + K, \tag{3-12}$$

in which $m_0 c^2$ is called the *rest energy* of the particle. The rest energy (by definition) is the energy of the particle at rest, when $u = 0$ and $K = 0$. The total energy of the particle (Eq. 3-12) is the sum of its rest energy and its kinetic energy.

The relativistic expression for $K$ must reduce to the classical result, $\frac{1}{2} m_0 u^2$, when $u/c \ll 1$. Let us check this. From Eq. 3-11b,

$$K = m_0 c^2 \left( \frac{1}{\sqrt{1 - u^2/c^2}} - 1 \right)$$

$$= m_0 c^2 \left[ \left( 1 - \frac{u^2}{c^2} \right)^{-1/2} - 1 \right],$$

and the binomial theorem expansion in $(u/c)$ gives

$$K = m_0 c^2 \left[ 1 + \frac{1}{2} \left( \frac{u}{c} \right)^2 + \frac{3}{8} \left( \frac{u}{c} \right)^4 + \cdots - 1 \right]$$

$$= \frac{1}{2} m_0 u^2 \left[ 1 + \frac{3}{4} \left( \frac{u}{c} \right)^2 + \cdots \right].$$

(3-11d)

We see that, as $u \to 0$, the second term in square brackets above become negligible in comparison to the first term and the expression for $K$ approaches the classical value, $\frac{1}{2} m_0 u^2$, thereby confirming the Newtonian limit of the relativistic result.

It is interesting to notice also that, as $u \to c$ in Eq. 3-11b, the kinetic energy $K$ tends to infinity. That is, from Eq. 3-10, an infinite amount of work would need to be done on the particle to accelerate it up to the speed of light. Once again we find $c$ playing the role of a limiting velocity. Note also from Eq. 3-11a, $K = (m - m_0)c^2$, that a change in the kinetic energy of a particle is related to a change in its relativistic mass.

## EXAMPLE 3. _____

*A Special Speed.* What is the speed of a particle whose kinetic energy $K$ is equal to its rest energy $m_0 c^2$?

The kinetic energy of a particle is defined from Eq. 3-11a, or

$$K = (m - m_0)c^2.$$

If we substitute $m_0 c^2$ for $K$ in this equation we find, after cancellation and minor rearrangement, that

$$m = 2m_0,$$

in which $m$ is the relativistic mass of the particle. Substituting for $m$ from Eq. 3-4b gives us

$$\frac{m_0}{\sqrt{1 - \beta^2}} = 2m_0,$$

or, cancelling and rearranging,

$$2\sqrt{1 - \beta^2} = 1.$$

Solving yields $\beta = 0.866$, and thus $u = \beta c = 0.866c$ is the speed of a particle whose kinetic energy is equal to its rest energy. Because the rest mass $m_0$ of the particle cancelled out in our derivation, the nature of the particle does not matter. The same speed holds for all particles, be they electrons, protons, or—for that matter—baseballs.

_____

We often seek a connection between the kinetic energy $K$ of a rapidly moving particle and its momentum $p$. This can be found by eliminating $u$ between Eq. 3-11b and Eq. 3-5. You can verify that the result is

$$(K + m_0 c^2)^2 = (pc)^2 + (m_0 c^2)^2,$$

(3-13a)

which, with the total energy $E = K + m_0 c^2$, can also be written as

$$E^2 = (pc)^2 + (m_0 c^2)^2.$$

(3-13b)

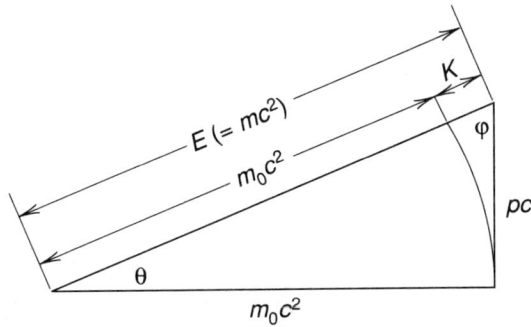

**FIGURE 3-3.** A mnemonic device, using a right triangle and the Pythagorean relation, to help in remembering the relations between total energy $E$, rest energy $m_0c^2$, and momentum $p$; see Eq. 3-13b. Shown also is the relation $E = m_0c^2 + K$ between total energy, rest energy, and kinetic energy. You can show that $\sin \theta = \beta$ and $\sin \phi = 1/\gamma$.

The right triangle of Fig. 3-3 is a useful mnemonic device for remembering Eq. 3-13.

The relationship between $K$ and $p$ (Eq. 3-13a) should reduce to the Newtonian expression $p = \sqrt{2m_0K}$ for $u/c \ll 1$. To see that it does, let us expand Eq. 3-13a, obtaining

$$K^2 + 2Km_0c^2 = p^2c^2. \qquad (3\text{-}13c)$$

When $u/c \ll 1$, the kinetic energy $K$ of a moving particle will always be much less than its rest energy $m_0c^2$. Under these circumstances, the first term on the left above $(K^2)$ can be neglected in comparison with the second term $(2Km_0c^2)$, and the equation becomes $p = \sqrt{2m_0K}$, as required.

The relativistic expression, Eq. 3-13b, often written as

$$E = c\sqrt{p^2 + m_0^2c^2}, \qquad (3\text{-}14)$$

is useful in high-energy physics to calculate the total energy of a particle when its momentum is given, or vice versa. By differentiating Eq. 3-14 with respect to $p$, we can obtain another useful relation:

$$\frac{dE}{dp} = \frac{pc}{\sqrt{m_0^2c^2 + p^2}} = \frac{pc^2}{c\sqrt{m_0^2c^2 + p^2}} = \frac{pc^2}{E}.$$

But with $E = mc^2$ and $p = mu$ this reduces to

$$\frac{dE}{dp} = u, \qquad (3\text{-}15)$$

a result that, incidentally, is also valid in classical dynamics.

**EXAMPLE 4.** _____

***Approximations at Low Speeds.*** Show that when $u/c < 0.1$, then (a) the ratio $K/m_0c^2$ is less than $1/200$ ($= 0.005$) and the classical expressions for (b) the kinetic energy and (c) the mo-

mentum of a particle may be used with an error of less than 1 percent.

(a) The kinetic energy of a particle is given by Eq. 3-11b, or

$$K = m_0 c^2 \left( \frac{1}{\sqrt{1 - \beta^2}} - 1 \right).$$

or, for $\beta = 0.1$,

$$\frac{K}{m_0 c^2} = \frac{1}{\sqrt{1 - (0.1)^2}} - 1$$

$$= 1.0050 - 1 = 0.0050.$$

If $\beta < 0.1$, it is easy to show that this ratio is correspondingly less than 0.0050.

(b) Now let us investigate the accuracy of the classical kinetic energy relationship ($K_c = \frac{1}{2} m_0 u^2$) for $\beta < 0.1$. Using the series expansion of Eq. 3-11d, we can write for $K$, the relativistic expression for the kinetic energy,

$$K = K_c (1 + \tfrac{3}{4} \beta^2 + \cdots).$$

For $e$, the percent error involved in our choice of formula, we then have, keeping only the first two terms in the expression for $K$,

$$e = \frac{K - K_c}{K} \times 100$$

$$= \frac{K_c (1 + \tfrac{3}{4} \beta^2) - K_c}{K_c (1 + \tfrac{3}{4} \beta^2)} \times 100$$

$$= \frac{300 \beta^2}{4 + 3 \beta^2}.$$

For $\beta = 0.1$ this yields

$$e = \frac{(300)(0.1)^2}{4 + (3)(0.1)^2} = 0.7\%,$$

which is less than 1 percent.

(c) We now turn our attention to the classical momentum formula, $p_c = m_0 u$. The relativistic formula for the momentum follows from Eq. 3-4c as

$$p = mu = \gamma m_0 u = \gamma p_c.$$

For $e$, the percent error found by using the classical formula, we have

$$e = \frac{p - p_c}{p} \times 100$$

$$= \frac{\gamma p_c - p_c}{\gamma p_c} \times 100$$

$$= 100 \left( \frac{\gamma - 1}{\gamma} \right).$$

For $\beta = 0.1$ we can easily show that $\gamma (= 1/\sqrt{1 - \beta^2}) = 1.0050$. Thus,

$$e = 100 \left( \frac{1.0050 - 1}{1.0050} \right) = 0.5\%,$$

which is less than 1 percent.

---

**EXAMPLE 5.** _____

***Approximations at High Speeds.*** **Show that when** $u/c > 99/100 (= 0.99)$, then (a) the ratio $K/m_0 c^2$ is greater than 6 and (b) the relativistic relation $p' = E/c$ for a zero-rest-mass particle may be used, with an error of less than 1 percent, for the momentum of a particle whose rest mass is actually $m_0$.

(a) The kinetic energy of a particle is given by Eq. 3-11c, or

$$K = m_0 c^2 (\gamma - 1).$$

The Lorentz factor $\gamma$ for $\beta = 0.99$ is readily found from

$$\gamma = \frac{1}{\sqrt{1 - \beta^2}} - \frac{1}{\sqrt{1 - (0.99)^2}} = 7.09.$$

Thus we have

$$K/m_0 c^2 = \gamma - 1 = 7.09 - 1 = 6.09,$$

which is greater than 6. If $\beta > 0.99$, it is easy to

show that this ratio will be correspondingly greater than 6.09.

(b) The total energy $E$ of a particle is equal to $mc^2$ in which $m$ is the relativistic mass. From $p' = E/c$, we then have

$$p' = \frac{E}{c} = \frac{mc^2}{c} = mc,$$

whereas the formula for the momentum of a particle whose rest mass is *not* zero is simply (see Eq. 3-5) $p = mu$. Thus $e$, the percent error involved in using the approximate formula, is

$$e = \frac{p - p'}{p} \times 100$$

$$= \frac{mu - mc}{mu} \times 100$$

$$= 100 \left( \frac{\beta - 1}{\beta} \right).$$

For $\beta = 0.99$ this yields

$$e = 100 \left(\frac{0.99 - 1}{0.99}\right) = -1.01\%,$$

which is essentially 1 percent. It is easy to show that if $\beta > 0.99$, the error will be even smaller.

From this example and the preceding one we see that relativistic effects are small, though not necessarily negligible, when $u < 0.1c$; and that when $u > 0.99c$, purely relativistic effects predominate.

As a final consideration in the relativistic dynamics of a single particle, we look at the acceleration of a particle under the influence of a force. In general, the force is given by

$$\mathbf{F} = \frac{d\mathbf{p}}{dt} = \frac{d}{dt}(m\mathbf{u})$$

or

$$\mathbf{F} = m\frac{d\mathbf{u}}{dt} + \mathbf{u}\frac{dm}{dt}. \tag{3-16}$$

We know that $m = E/c^2$, so

$$\frac{dm}{dt} = \frac{1}{c^2}\frac{dE}{dt} = \frac{1}{c^2}\frac{d}{dt}(K + m_0c^2) = \frac{1}{c^2}\frac{dK}{dt}.$$

But

$$\frac{dK}{dt} = \frac{(\mathbf{F} \cdot d\mathbf{l})}{dt} = \mathbf{F} \cdot \frac{d\mathbf{l}}{dt} = \mathbf{F} \cdot \mathbf{u},$$

so

$$\frac{dm}{dt} = \frac{1}{c^2}\mathbf{F} \cdot \mathbf{u}.$$

We can now substitute this into Eq. 3-16 and obtain

$$\mathbf{F} = m\frac{d\mathbf{u}}{dt} + \frac{\mathbf{u}(\mathbf{F} \cdot \mathbf{u})}{c^2}.$$

The acceleration $\mathbf{a}$ is defined by $\mathbf{a} = d\mathbf{u}/dt$, so the general expression for acceleration is

$$\mathbf{a} = \frac{d\mathbf{u}}{dt} = \frac{\mathbf{F}}{m} - \frac{\mathbf{u}}{mc^2}(\mathbf{F} \cdot \mathbf{u}). \tag{3-17}$$

What this equation tells us at once is that, in general, the acceleration $\mathbf{a}$ is *not* parallel to the force in relativity, since the last term above is in the direction of the velocity $\mathbf{u}$.

## EXAMPLE 6.

*Accelerating Electrons.* A certain linear accelerator is accelerating electrons by letting them fall through a potential difference of 4.50 MV. Find (*a*) the kinetic energy, (*b*) the mass, and (*c*) the speed of the electrons as they emerge from the accelerator into the laboratory.

(*a*) The charge on the electron is $-e$, where $e(= 1.60 \times 10^{-19}$ C) is the electronic charge. The kinetic energy is given by

$$K = q(V_i - V_f)$$
$$= (-1.00 \text{ electronic charge})$$
$$\times (0 - 4.50 \times 10^6 \text{ V})$$
$$= 4.50 \text{ MeV}.$$

In SI units we have, for this same quantity,

$$K = (-1.60 \times 10^{-19} \text{ C})(0 - 4.50 \times 10^6 \text{ V})$$
$$= 7.20 \times 10^{-13} \text{ J}.$$

(b) We can rearrange Eq. 3-11a $[K = (m - m_0)c^2]$ in the form

$$\frac{m}{m_0} = \frac{K}{m_0c^2} + 1$$

$$= \frac{(7.20 \times 10^{-13} \text{ J})}{(9.11 \times 10^{-31} \text{ kg})(3.00 \times 10^8 \text{ m/s})^2} + 1$$

$$= 8.78 + 1 = 9.78.$$

Thus the relativistic mass $m$ of such accelerated electrons is almost 10 times their rest mass. We can find $m$ from

$$m = 9.78m_0 = (9.78)(9.11 \times 10^{-31} \text{ kg})$$

$$= 8.91 \times 10^{-30} \text{ kg}.$$

(c) Finally, we can rearrange Eq. 3-4b $(m = m_0/\sqrt{1 - \beta^2})$ to read

$$\beta = \sqrt{1 - \left(\frac{m_0}{m}\right)^2}$$

$$= \sqrt{1 - \left(\frac{1}{9.78}\right)^2}$$

$$= 0.9948.$$

Thus,

$$u = \beta c = (0.9948)(3.00 \times 10^8 \text{ m/s})$$

$$= 2.98 \times 10^8 \text{ m/s}.$$

These electrons are moving very close indeed to the speed of light. At such speeds (see Problem 5), a relatively small fractional increase in speed corresponds to a sizable fractional increase in kinetic energy.

The calculations of this example illustrate the point made in Example 5, namely, that if $\beta > 0.99$, then relativistic effects are central to the situation and, in particular, the kinetic energy $K$ will be at least six times greater than the rest energy $m_0c^2$. In part (c) we see that $\beta$ does indeed exceed 0.99; by inspection of the calculation in (b), we see that $K/m_0c^2$ is 8.78, which is greater than 6.

## EXAMPLE 7.

### A Charged Particle Moving in a Magnetic Field.

(a) Show that, in a region in which there is a uniform magnetic field, a charged particle entering at right angles to the field moves in a circle whose radius is proportional to the particle's momentum.

The force on the particle is

$$\mathbf{F} = q\mathbf{u} \times \mathbf{B},$$

which is at right angles both to $\mathbf{u}$ and to $\mathbf{B}$. The scalar product $\mathbf{F} \cdot \mathbf{u}$ in Eq. 3-17 is thus zero and the acceleration, which is now entirely in the direction of the force, is given by

$$\mathbf{a} = \frac{\mathbf{F}}{m} = \frac{q}{m} \mathbf{u} \times \mathbf{B}.$$

Because the acceleration is always at right angles to the particle's velocity $\mathbf{u}$, the speed of the particle is constant and the particle moves in a circle. Let the radius of the circle be $r$, so that the centripetal acceleration is $u^2/r$. We equate this to the acceleration obtained from above, $a = quB/m$, and find

$$\frac{quB}{m} = \frac{u^2}{r}$$

or

$$r = \frac{mu}{qB} = \frac{p}{qB}. \qquad (3\text{-}18)$$

Hence, the radius is proportional to the momentum $p (= mu)$.

Notice that both the equation for the acceleration and the equation for the radius (Eq. 3-18) are identical in form to the classical results, but that the rest mass $m_0$ of the classical formula is replaced by the relativistic mass $m = m_0/\sqrt{1 - u^2/c^2}$.

How would the motion change if the initial velocity of the charged particle had a component parallel to the magnetic field?

(b) Compute the radius, both classically and relativistically, of a path of a 10-MeV electron moving at right angles to a uniform magnetic field of strength 2.0 T.

*Classically,* we have $r = m_0 u / qB$. The classical relation between kinetic energy and momentum is $p = \sqrt{2 m_0 K}$, so

$$p = \sqrt{2 m_0 K}$$
$$= \sqrt{2(9.11 \times 10^{-31} \text{ kg})(10 \text{ MeV})(1.60 \times 10^{-13} \text{ J/MeV})}$$
$$= 1.7 \times 10^{-21} \text{ kg} \cdot \text{m/s}.$$

Then

$$r = \frac{m_0 u}{qB} = \frac{p}{qB} = \frac{1.7 \times 10^{-21} \text{ kg} \cdot \text{m/s}}{(1.60 \times 10^{-19} \text{ C})(2.0 \text{ T})}$$
$$= 5.3 \times 10^{-3} \text{ m} = 5.3 \text{ mm}.$$

*Relativistically,* we have $r = mu/qB$. The relativistic relation between kinetic energy and momentum (Eq. 3-13a) may be written as

$$p = \frac{1}{c} \sqrt{(K + m_0 c^2)^2 - (m_0 c^2)^2}.$$

Here, the rest energy of an electron, $m_0 c^2$, equals 0.511 MeV, so that

$$p = \frac{1}{3.0 \times 10^8} \sqrt{(10 + 0.511)^2 - (0.511)^2}$$
$$\times \frac{\text{MeV} \cdot \text{s}}{\text{m}} \times (1.60 \times 10^{-13} \text{ J/MeV})$$
$$= 5.6 \times 10^{-21} \text{ kg} \cdot \text{m/s}.$$

Then

$$r = \frac{mu}{qB} = \frac{p}{qB} = \frac{5.6 \times 10^{-21} \text{ kg} \cdot \text{m/s}}{(1.60 \times 10^{-19} \text{ C})(2.0 \text{ T})}$$
$$= 1.8 \times 10^{-2} \text{ m} = 18 \text{ mm},$$

which is substantially larger than the classical prediction. Experiment supports the relativistic prediction conclusively.

## 3-5 Some Experimental Results*

Early (1909) experiments in relativistic dynamics by Bucherer made use of Eq. 3-18. Electrons (from the $\alpha$ decay of radioactive particles) enter a velocity selector, which determines the speed of those that emerge, and then enter a uniform magnetic field, where the radius of their circular path can be measured. Bucherer's results are shown in Table 3-1.

The first column gives the measured speeds in terms of the fraction of the speed of light. The second column gives the ratio $e/m$ computed from the measured quantities in Eq. 3-18 as $e/m = u/rB$. It is clear that the value of $e/m$ varies with the speed of the electrons. The third column gives the calculated values of

**TABLE 3-1**
Bucherer's Results

| $\beta$ $(= u/c)$ (Measured) | $e/m$ $(= u/rB)$ $10^{11}$ C/kg (Measured) | $e/m_0$ $(= e/m\sqrt{1 - \beta^2})$ $10^{11}$ C/kg (Calculated) |
|---|---|---|
| 0.32 | 1.66 | 1.75 |
| 0.38 | 1.63 | 1.76 |
| 0.43 | 1.59 | 1.76 |
| 0.52 | 1.51 | 1.76 |
| 0.69 | 1.28 | 1.77 |

*See also reference 8.

$e/m \sqrt{1 - \beta^2} = e/m_0$, which are seen to be constant. The results are consistent with the relativistic relation

$$r = \frac{m_0 u}{qB\sqrt{1 - \beta^2}}$$

rather than the classical relation $r = m_0 u/qB$ and can be interpreted as confirming Eq. 3-4b, $m = m_0/\sqrt{1 - \beta^2}$, for the variation of mass with speed. Many similar experiments have since been performed, greatly extending the range of $u/c$ and always resulting in confirmation of the relativistic results (see Fig. 3-4).

You may properly ask why, in measuring a variation of $e/m$ with speed, we attribute the variation solely to the mass rather than to the charge, for instance, or some other more complicated effect. We might have concluded, for example, that $e = e_0 \sqrt{1 - \beta^2}$. Actually, we have implicitly assumed above that the charge on the electron is independent of its speed. This assumption is a direct consequence of relativistic electrodynamics, wherein the charge of a particle is not changed by its motion. That is, charge is an invariant quantity in relativity. This is plausible, as a little thought shows, for otherwise the neutral character of an atom would be upset merely by the motion of the electrons in it. As a clincher, of course, we turn to experiment; we then find that experiment not only verifies relativity theory as a whole, but also confirms directly this specific result of the constancy of $e$ (see [1] for an analysis of such an experiment). Beyond this, experiments with neutrons—neutral particles not involving electric charge—demonstrate directly the same variation of mass with velocity as for charged particles.

An experiment in relativistic dynamics was also carried out by Bertozzi [2]. In this experiment electrons are accelerated to high speed in the electric field of a linear accelerator and emerge into a vacuum chamber. Their speed can be measured by

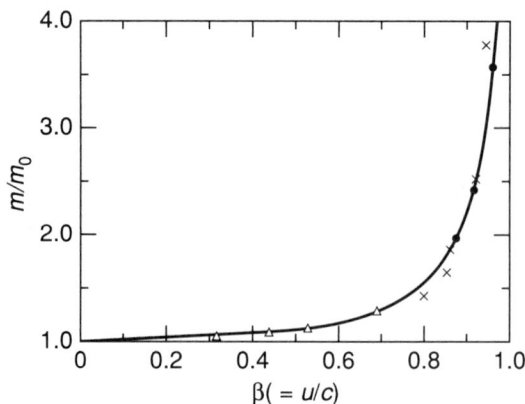

**FIGURE 3-4.** Some experimental measurements of the ratio $m/m_0$ for the electron, at various speed parameters $\beta(= u/c)$. The symbols stand for the work of Kaufmann ($\times$, 1901), Bucherer ($\triangle$, 1908), and Bertozzi ($\bullet$, 1964).

determining the time of flight in passing two targets of known separation. As we vary the voltage of the accelerator, we can plot the values of $eV$, the kinetic energy of the emerging electrons, versus the measured speed $u$. In the experiment, an independent check was made to confirm the relation $K = eV$. This is accomplished by stopping the electrons in a collector, where the kinetic energy of the absorbed electrons is converted into heat energy, which raises the temperature of the collector, and determining the energy released per electron by calorimetry. It is found that the average kinetic energy per electron before impact, measured in this way, agrees with the kinetic energy obtained from $eV$.

Figure 3-5 shows the results of this experiment. We see that the measured values of $K$ agree very well with the relativistic prediction but not at all with the classical prediction. Note also that in all cases the measured value of $u$ is less than $c$ or, what is the same thing, the value of $\beta$ is less than unity. This is further direct confirmation of the role of $c$ as a limiting speed. We see that, as $u \rightarrow c$, small changes in $u$ correspond to increasingly large changes in the kinetic energy $K$. To attain a given speed we always need more kinetic energy than is classically predicted; by extrapolation, we would need infinite energy to accelerate an electron to the speed of light.

Note carefully that the relativistic formula for kinetic energy is *not* $\frac{1}{2}mu^2$; this shows the danger in assuming that we can simply substitute the relativistic mass for the rest mass in generalizing a classical formula to a relativistic one. This is not so for the kinetic energy.

Although the experimental checks of relativistic dynamics that we have described are direct and to the point, perhaps the most convincing evidence that relativistic

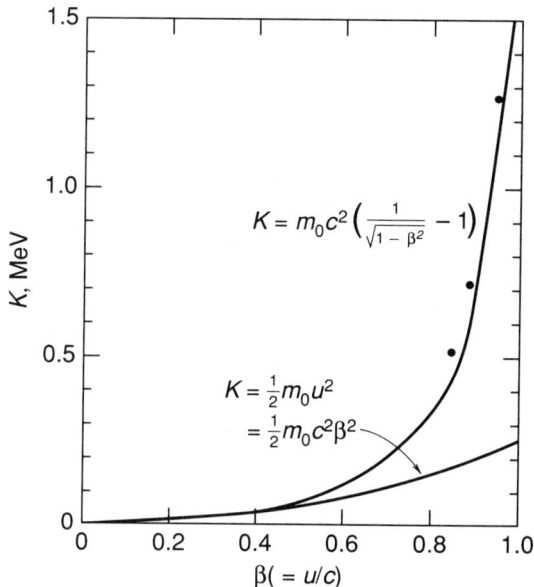

**FIGURE 3-5.** Bertozzi's experimental points (●) are seen to fit the relativistic expression for the kinetic energy of an electron at various speeds, rather than the classical expression.

dynamics is valid lies in the design and successful operation of high-energy particle accelerators (see *Physics,* Part II Sect. 33-7). The proton synchrotron [3] at the Fermi National Accelerator Laboratory (Fermilab) at Batavia, Illinois, is a prime example; see Fig. 3-6. It is a multimillion-dollar device whose design is based squarely on the relativistic predictions that (1) the speed of light is a limiting speed for material particles and (2) the mass of a particle increases with velocity as described by Eq. 3-4. If these predictions were not correct, this huge accelerator, built around a ring-shaped magnet ~4 miles in circumference, simply would not work.

Consider a proton moving with speed $u$ at right angles to a uniform magnetic field $B$. As we have seen, it will move in a circular path whose radius $R$ is given by Eq. 3-18, or $R = mu/Be$, in which $m$ is the relativistic mass. The angular frequency $\omega$ at which the proton circulates in this field is called the proton's *cyclotron frequency* in that field and is given by

$$\omega = \frac{u}{R} = \frac{Be}{m}.$$

(3-19)

The circulating protons can be accelerated by allowing them to pass repeatedly through a small region in which an alternating electric field is established, the field

**FIGURE 3-6.**  The proton synchrotron at Fermilab; a view through the tunnel. (Courtesy of Fermilab)

giving the proton an accelerating impulse or "kick" on every passage. It would be simplest to provide these accelerating impulses from a fixed-frequency oscillator. However, there is the complication that $\omega$, the cyclotron frequency of the circulating protons, is *not* constant but, as Eq. 3-19 shows, decreases with time throughout the accelerating cycle because of the relativistic increase of mass with speed. This can be compensated for, however, by adjusting the magnetic field $B$ in Eq. 3-19 in a cyclic way so that $\omega$ does indeed remain constant and thus in resonance at all times with an oscillator of fixed frequency.

There is a problem, though. Under the scheme just outlined, the orbit radius, given by (see Eq. 3-19)

$$R = \frac{u}{\omega},$$    (3-20)

would increase as the speed increases. Now, it is highly desirable to maintain the orbit radius constant during the acceleration process, because the steering magnets can then be ring-shaped, resulting in great cost savings. Inspection of Eq. 3-20 shows that $R$ can indeed be held constant if the oscillator frequency (which must always be equal to the cyclotron frequency $\omega$) is *also* increased during the accelerating cycle, to compensate for the increase in $u$.

Table 3-2 shows some characteristics of the proton synchrotron at the Fermi National Laboratory.

**TABLE 3-2**
The Proton Synchrotron at Fermilab

| | |
|---|---|
| Maximum proton energy | 500 GeV |
| Repetition rate | 5 pulses/min |
| Internal beam intensity | $6 \times 10^{12}$ protons/s |
| Radius of the ring | 1000 m |
| Radius of curvature of the proton path | 750 m |
| Steel weight | 9000 tons |
| Copper weight (in magnet coils) | 850 tons |
| Magnetic field at injection | 40 mT |
| Magnetic field at 300 GeV | 1.35 T |
| Mean power to magnets | 36 MW |
| Oscillator range | 53.08–53.10 MHz |

**EXAMPLE 8.**

*The Fermilab Accelerator.* The proton synchrotron at Fermilab is accelerating protons to a kinetic energy of 500 GeV. At this energy, find (a) the ratio $m/m_0$, (b) the speed parameter $\beta$, (c) the magnetic field $B$ at the orbit position, and (d) the period $T$ of the circulating 500-GeV protons. (See Table 3-2 for other needed data.)

(a) From Eq. 3-4c we see that the ratio $m/m_0$ is simply the Lorentz factor $\gamma$. From Eq.

3-11c $[K = (\gamma - 1)m_0c^2]$, we can write

$$\gamma = \frac{K}{m_0c^2} + 1 = \frac{500 \times 10^3 \text{ MeV}}{938 \text{ MeV}} + 1$$

$$= 532 + 1 = 533.$$

Thus the relativistic mass of a 500-GeV proton is 533 times its rest mass.

(b) Knowing $\gamma$, we can solve Eq. 2-10 for the speed parameter $\beta$, obtaining

$$\beta = \sqrt{1 - \left(\frac{1}{\gamma}\right)^2} = \sqrt{1 - \left(\frac{1}{533}\right)^2}$$

$$= 0.9999982.$$

(c) The magnetic field follows from Eq. 3-18,
or

$$B = \frac{mu}{eR} = \frac{(\gamma m_0)(\beta c)}{eR} = \frac{\gamma \beta m_0 c}{eR}$$

$$= \frac{(533)(1)(1.67 \times 10^{-27} \text{ kg})(3.00 \times 10^8 \text{ m/s})}{(1.60 \times 10^{-19} \text{ C})(750 \text{ m})}$$

$$= 2.23 \text{ T.}$$

This must be the value of $B$ at the end of the accelerating cycle for 500-GeV protons. The value of $R$ used to calculate $B$ must then be the actual radius of curvature of the proton path, not the radius of the magnet ring; these differ because the "ring" contains a number of straight segments (see Table 3-2).

(d) At 500 GeV, the proton speed is virtually $c$ and the period is given by

$$T = \frac{2\pi R'}{u} = \frac{(2\pi)(1000 \text{ m})}{(3.00 \times 10^8 \text{ m/s})} = 20.9 \text{ μs.}$$

Note that the magnet ring used in *this* calculation is the actual effective radius of the magnetic ring; see Table 3-2.

## 3-6 The Equivalence of Mass and Energy

In Section 3-3 we found that the laws of the conservation of momentum and of mass could be preserved by generalizing the classical definition of mass to a relativistic mass $m$; see Eqs. 3-4. In Section 3-4 we introduced the concept of the total energy $E$ of a single particle, defining it equivalently as

$$E = mc^2 \qquad (3\text{-}21a)$$

or

$$E = m_0 c^2 + K, \qquad (3\text{-}21b)$$

in which $m_0 c^2$ is the rest energy and $K$ is the kinetic energy of the (single) particle. In this section we seek to extend the concept of total energy from single particles to systems of interacting particles.

Let us return to our examination of the totally inelastic collision of two identical particles, described in Fig. 3-2. Equation 3-21a suggests that, if relativistic mass is conserved in such a collision (and we have shown that it is), then the total energy $E$ of the system of particles will also be conserved. Equation 3-21b suggests further that, if $E$ is conserved for this system of particles, then any net change in the kinetic energies of the interacting particles must be balanced by an equal but opposite net change in the rest energies of those particles. Let us verify that these energy exchanges do in fact occur. We shall be led from this study to important conclusions about the nature of energy conservation in relativity and about the equivalence of mass and energy.

Conservation of relativistic mass in the $S$ frame of Fig. 3-2 (see Example 2) requires that

$$M_0 = 2m.$$

We can substitute for the relativistic mass $m$ from Eq. 3-4b, obtaining

$$M_0 = \frac{2m_0}{\sqrt{1 - \beta^2}} = 2m_0\gamma,$$

which shows at once that, although relativistic mass is conserved in this collision, *rest mass* is not; $M_0$, the rest mass after the collision, is greater than $2m_0$, the rest mass before the collision. Thus we can write, for the increase in rest mass during the collision,

$$\text{increase in rest mass} = M_0 - 2m_0 = 2m_0\gamma - 2m_0$$
$$= 2m_0(\gamma - 1). \qquad (3\text{-}22a)$$

Now let us look at the kinetic energy situation. In frame $S$ of Fig. 3-2$a$ the individual particles have a combined kinetic energy before the collision given by Eq. 3-11$c$ as $2m_0(\gamma - 1)c^2$. After the collision, no kinetic energy remains. In place of the "lost" kinetic energy there appears internal (thermal) energy, recognizable by the rise in temperature of the colliding particles. Thus we can write

$$\text{increase in internal energy} = \text{decrease in kinetic energy}$$
$$= 2m_0(\gamma - 1)c^2. \qquad (3\text{-}22b)$$

Comparison of these last two equations allows us to write, for this collision at least,

$$(\text{decrease in kinetic energy}) = (\text{increase in rest mass})(c^2),$$

or equivalently

$$(\text{increase in thermal energy}) = (\text{increase in rest mass})(c^2).$$

Thus we see that the decrease in kinetic energy for this isolated system is balanced by a corresponding increase in rest energy, just as we expected. We see also that the thermal energy that appears in the system is associated with an increase in the rest mass of the system. Hence rest mass is equivalent to energy (rest-mass energy) and must be included in applying the conservation of energy principle.

All of the foregoing justifies our making a great extrapolation from the simple special case we have examined and asserting a general principle, namely, that mass and energy are two aspects of a single invariant quantity, which we can call *mass-energy*. We can find the energy equivalent of a given mass by multiplying by $c^2$. In the same way, we can find the mass equivalent of a given amount of energy by dividing by $c^2$.

The relation

$$E = mc^2 \qquad (3\text{-}23)$$

expresses the fact that mass-energy can be expressed in energy units ($E$) or equivalently in mass units ($m = E/c^2$). In fact, it has become common practice to refer to masses in terms of electron volts, such as saying that the rest mass of an electron is 0.511 MeV, for convenience in energy calculations. Likewise, particles of zero rest mass (such as photons, see below) may be assigned an effective mass equivalent to their energy. Indeed, the mass that we associate with various forms of energy really has all the properties we have heretofore given to mass, properties such as inertia, weight, contribution to the location of the center of mass of a system, and so forth. We shall exhibit some of these properties later in the chapter (see also Ref. 4).

Equation 3-23, $E = mc^2$, is, of course, one of the famous equations of physics. It has been confirmed by numerous practical applications and theoretical consequences. Einstein, who derived the result originally in another context, made the bold hypothesis that it was universally applicable. He considered it to be the most significant consequence of his special theory of relativity.

Thus we see that the conclusions we have drawn here from our study of two-body collisions are perfectly consistent with the single-particle equations we developed in Section 3-4. Consider Eq. 3-11*a:*

$$mc^2 = m_0c^2 + K.$$

We can differentiate this result, obtaining

$$\frac{dK}{dt} = c^2 \frac{dm}{dt} , \tag{3-24}$$

which states that a change in the kinetic energy of a particle causes a proportionate change in its (relativistic) mass. That is, mass and energy are equivalent, their units differing by a factor $c^2$.

If the kinetic energy of a body is regarded as a form of external energy, then the rest-mass energy may be regarded as the energy of the body. This internal energy consists, in part, of such things as molecular motion, which changes when heat energy is absorbed or given up by the body, or intermolecular potential energy, which changes when chemical reactions (such as dissociation or recombination) take place. Or the internal energy can take the form of atomic potential energy, which can change when an atom absorbs radiation and becomes excited or emits radiation and is de-excited, or nuclear potential energy, which can be changed by nuclear reactions. The largest contribution to the internal energy is, however, the total rest-mass energy contributed by the "fundamental" particles, which is regarded as the primary source of internal energy. This too, may change, as, for example, in electron-positron creation and annihilation. The rest mass (or proper mass) of a body, therefore, is not a constant, in general. Of course, if there are no changes in the internal energy of a body (or if we consider a closed system through which energy is not transferred), then we may regard the rest mass of the body (or of the system) as constant.

This view of the internal energy of a particle as equivalent to rest mass suggests an extension to a collection of particles. We sometimes regard an atom as a particle and assign it a rest mass, for example, although we know that the atom consists of many particles with various forms of internal energy. Likewise, we can assign a rest mass to any collection of particles in relative motion, in a frame in which the center of mass is at rest (that is, in which the resultant momentum is zero). The rest mass of the system as a whole would include the contributions of the internal energy of the system to the inertia.

Returning our attention now to collisions or interactions between bodies, we have seen that the total energy is conserved and that the conservation of total energy is equivalent to the conservation of (relativistic) mass. Although we showed this explicitly for a totally inelastic collision only, it holds regardless of the nature of the collision. To retain the conservation laws we must consider mass and energy as two different aspects of a single entity, namely, mass-energy. In the energy balance we must consider the masses of particles, and in the mass balance we must consider the energy of radiation, for example. The formula $E = mc^2$ can be regarded as giving the rate of exchange between two interchangeable currencies, $E$ and $m$.

In classical physics we had two separate conservation principles: (1) the conservation of (classical) mass, as in chemical reactions, and (2) the conservation of energy. In relativity, these merge into one conservation principle, that of conservation of mass-energy. The two classical laws may be viewed as special cases that would be expected to agree with experiment only if energy transfers into or out of the system are so small compared to the system's rest mass that the corresponding fractional change in rest mass of the system is too small to be measured.

## EXAMPLE 9.

*Pulling Apart a Deuteron.* A convenient mass unit to use when dealing with atoms or nuclei is the *atomic mass unit* (abbreviation u). It is defined so that the atomic mass of one atom of the most common carbon isotope (carbon-12) is exactly 12 such units and has the value 1 u = $1.66 \times 10^{-27}$ kg. The rest mass of the proton (the nucleus of a hydrogen atom) is 1.00728 u, and that of the neutron (a neutral particle and a constituent of all nuclei except hydrogen) is 1.00867 u. A deuteron (the nucleus of heavy hydrogen) is known to consist of a proton and a neutron and to have a rest mass of 2.01355 u. What energy is required to break up a deuteron into its constituent particles?

The rest mass of a deuteron is *less than* the combined rest masses of a proton and a neutron by

$$\Delta m_0 = [(1.00728 + 1.00867) - 2.01355] \, u$$
$$= 0.00240 \, u,$$

which is equivalent, in energy terms, to

$$\Delta m_0 c^2 = (0.00240 \, u)(1.66 \times 10^{-27} \, kg/u)$$
$$\times (3.00 \times 10^8 \, m/s)^2$$
$$= (3.59 \times 10^{-13} \, J)$$

$$\times (1 \, MeV/1.60 \times 10^{-13} \, J)$$
$$= 2.23 \, MeV.$$

When a proton and a neutron at rest combine to form a deuteron, this amount of energy is *given off* in the form of electromagnetic (gamma) radiation. If the deuteron is to be broken up into a proton and a neutron, this same amount of energy must be *added to* the deuteron.

Notice that

$$\frac{\Delta m_0}{M_0} = \frac{0.00240 \, u}{2.01355 \, u} = 0.12\%.$$

This fractional rest-mass change is characteristic of the magnitudes found in nuclear reactions.

In solving problems involving nuclear interactions, in which the energies are typically measured in MeV and the masses in atomic mass units, it is convenient to realize (see Problem 61) that $c^2$ can be written as

$$c^2 = 931 \, MeV/u.$$

In this example, then, we could simply have multiplied the mass change (= 0.00240 u) by $c^2$ written in this form and obtained at once for the energy (0.00240 u)(931 MeV/u) or 2.23 MeV.

## EXAMPLE 10.

*Pulling Apart a Hydrogen Atom.* The energy $E_b$ required to break a hydrogen atom apart into its constituents—a proton and an electron—is called its *binding energy.* Its measured value is 13.58 eV. The rest mass $M_0$ of a hydrogen atom is 1.00783 u. By how much does the rest mass of this atom change when it is ionized? Is the change an increase or a decrease?

From the relation $E = mc^2$, we can write

$$\Delta m_0 = \frac{E_b}{c^2}$$

$$= \frac{13.58 \, eV}{931 \times 10^6 \, eV/u} = 1.46 \times 10^{-8} \, u,$$

in which $931 \times 10^6$ eV/u, as we have seen in Example 9, is simply a convenient way of writing $c^2$. The change is an *increase* because energy must be *added to* the atom to ionize it. We also note that

$$\frac{\Delta m_0}{M_0} = \frac{1.46 \times 10^{-8} \, u}{1.00783 \, u} \cong 1.5 \times 10^{-8}$$
$$= 1.5 \times 10^{-6}\%.$$

Such a fractional change in rest mass is actually smaller than the experimental errors involved in measuring the rest masses themselves. Thus, in interactions involving atoms (including all chemical reactions), the changes in rest mass are too small to detect and the classical principle of the conservation of (rest) mass is practically correct. As Example 9 shows, this statement is not true for reactions involving the nuclei of atoms.

In a paper [5] entitled "Does the Inertia of a Body Depend upon its Energy Content," Einstein writes:

> If a body gives off the energy $L$ in the form of radiation, its mass diminishes by $L/c^2$. The fact that the energy withdrawn from the body becomes energy of radiation evidently makes no difference, so that we are led to the more general conclusion that the mass of a body is a measure of its energy content. . . . It is not impossible that with bodies whose energy-content is variable to a high degree (for example, with radium salts) the theory may be successfully put to the test. If the theory corresponds to the facts, radiation conveys inertia between the emitting and absorbing bodies.

Experiment has abundantly confirmed Einstein's theory.

Today, we call such a pulse of radiation a photon and may regard it as a particle of zero rest mass. The relation $p = E/c$, taken from classical electromagnetism, is consistent with the result of special relativity for particles of "zero rest mass" since, from Eq. 3-14, $E = c\sqrt{p^2 + m_0^2 c^2}$, we find that $p = E/c$ when $m_0 = 0$. This is also consistent with the fact that photons travel with the speed of light because, from the relation $E = mc^2 = m_0 c^2/\sqrt{1 - u^2/c^2}$, the energy $E$ would go to zero as $m_0 \rightarrow 0$ for $u < c$. In order to keep $E$ finite (neither zero nor infinite) as $m_0 \rightarrow 0$, we must let $u \rightarrow c$. Strictly speaking, however, the term zero rest mass is a bit misleading, because it is impossible to find a reference frame in which photons (or anything that travels at the speed of light) are at rest (see Question 7). However, if $m_0$ is determined from energy and momentum measurements as $m_0 = \sqrt{(E/c^2)^2 - (p/c)^2}$, then $m_0 = 0$ when (as for a photon*) $p = E/c$. The result, that a particle of zero rest mass can have a finite energy and momentum and that such particles must move at the speed of light, is also consistent with the meaning we have given to rest mass as internal energy. For if rest mass is internal energy, existing when a body is at rest, then a "body" without mass has no internal energy. Its energy is all external, involving motion through space. Now, if such a body moved at a speed less than $c$ in one reference frame, we could always find another reference frame in which it *is* at rest. But if it moves at a speed $c$ in one reference frame, it will move at this same speed $c$ in all reference frames. It is consistent with the Lorentz transformation, then, that a body of zero rest mass should move at the speed of light and be nowhere at rest.

---

*For students who are unfamiliar with the relation $p = E/c$, found in electromagnetism, the argument can be run in reverse. Start with the relativistic relation $E = m_0 c^2/\sqrt{1 - u^2/c^2}$. This implies that $E$ approaches infinity if $u = c$, unless $m_0 = 0$. Therefore photons, which by definition have $u = c$, must have $m_0 = 0$. Then, from $E = c(p^2 + m_0^2 c^2)^{1/2}$, it follows that photons must satisfy the relation $p = E/c$. That this same result is found independently in classical electromagnetism illustrates the consistency between relativity and classical electromagnetism.

## EXAMPLE 11.

**The Sun is Losing Mass.** The earth receives radiant energy from the sun at the rate of 1340 W/m². At what rate is the sun losing rest mass because of its radiation? The sun's mass is now about $2.0 \times 10^{30}$ kg.

If we assume that the sun radiates uniformly in all directions, we can calculate its luminosity ($= dE/dt$) from

$$\frac{dE}{dt} = (1340 \text{ W/m}^2)(4\pi R^2)$$

$$= (1340 \text{ W/m}^2)(4\pi)(1.50 \times 10^{11} \text{ m})^2$$

$$= 3.79 \times 10^{26} \text{ W},$$

in which $R$ is the mean earth–sun distance.

The rate of mass loss is then

$$\frac{dm}{dt} = \frac{dE/dt}{c^2}$$

$$= \frac{3.79 \times 10^{26} \text{ J/s}}{(3.00 \times 10^8 \text{ m/s})^2}$$

$$= 4.21 \times 10^9 \text{ kg/s}.$$

At this rate, the fractional rate of loss of solar mass is

$$f = \frac{(4.21 \times 10^9 \text{ kg/s})(3.16 \times 10^7 \text{ s/y})}{2.0 \times 10^{30} \text{ kg}}$$

$$= 6.7 \times 10^{-14} \text{ y}^{-1}.$$

From this it can be shown that, if the sun had maintained its present luminosity since its formation about $5 \times 10^9$ y ago, it would have lost only about 0.03 percent of its rest mass.

## EXAMPLE 12.

**Another Approach to $E = mc^2$.** Here we present an "elementary derivation of the equivalence of mass and energy" attributable to Einstein. Consider a body $B$ at rest in frame $S$ (Fig. 3-7a). It emits simultaneously two pulses of radiation, each of energy $E/2$, one in the $+y$ direction and one in the $-y$ direction. The energy of $B$ therefore decreases by an amount $E$ in the emission process; from symmetry considerations, $B$ must remain at rest both during and after this process.

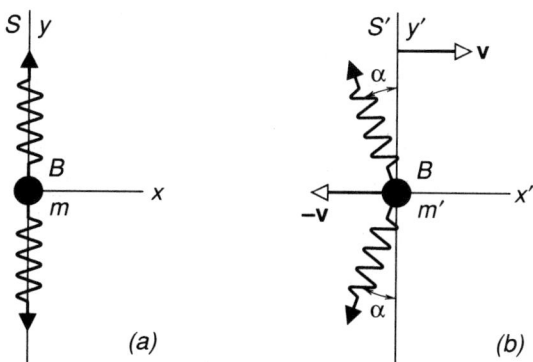

(a)

(b)

**FIGURE 3-7.** (a) Body $B$, at rest in frame $S$, emits two light pulses in opposite directions. (b) The same phenomenon as observed from frame $S'$. The light beams are deflected through an aberration angle $\alpha$.

Now consider these same events as viewed from frame $S'$, which is moving in the $+x$ direction with speed $v$ (assumed $\ll c$; see Fig. 3-7b). In this frame, body $B$ is seen to move in the $-x'$ direction with speed $v$. The two pulses of radiation, because of the aberration effect, now make a small angle $\alpha$ with the vertical ($y'$) axis. It is central to the proof to note that, because the state of motion of $B$ was not changed by the emission process in frame $S$, it cannot be changed by this process in frame $S'$. Thus the velocity of $B$ in frame $S'$ must remain $-\mathbf{v}$ after the two pulses have been emitted. If it were otherwise, there would be a violation of the principle of relativity, namely, that the laws of physics must be the same in all inertial reference frames.

Let us apply the law of conservation of momentum in frame $S'$. We assume that $v \ll c$, so we can use the classical form of this law. Before doing so we must point out that radiation has momentum, the amount being equal to the associated energy divided by $c$, the speed of light. Thus, each of our light pulses has a momentum, in its direction of motion, of magnitude $E/2c$. The assigning of momentum to radiation is a result from the classical theory of electromagnetism.

Before the emission process the momentum of the system in frame $S'$ is $-mv$, associated entirely with the body. After the emission process it is

$$-m'v - 2\left(\frac{E}{2c}\right)\sin\alpha.$$

Setting the momentum of the system before emission equal to its value after emission leads to

$$mv = m'v + \left(\frac{E}{c}\right)\sin\alpha.$$

Because $v \ll c$ the aberration angle $\alpha$ is very small, and we can replace $\sin\alpha$ by $\alpha$ in the above without appreciable error. Also (see Eq. 1-11),

classical aberration theory predicts that (again for $v \ll c$) $\alpha \cong v/c$. Replacing $\sin\alpha$ in the above equation by $v/c$ gives

$$m - m' = \frac{E}{c^2}.$$

Thus we see that body $B$, at rest in frame $S$, loses energy $E$ by emitting two light pulses and remains at rest, its (rest) mass decreased by $E/c^2$. Generalizing from this allows us to put

$$E = (\Delta m)c^2$$

as a relation valid for any process in which energy $E$ is emitted from or absorbed by a body.

## 3-7 Relativity and Electromagnetism

In the earlier sections we investigated the dynamics of a single particle using the relativistic equation of motion that was found to be in agreement with experiment for the motion of high-speed charged particles. There we introduced the relativistic mass and the total energy, including the rest-mass energy. However, all the formulas we used were applicable in one reference frame, which we called the laboratory frame. Often, as when analyzing nuclear reactions, it is useful to be able to transform these relations to other inertial reference frames, such as the center-of-mass frame. In such cases one uses the equations that connect the values of the momentum, energy, mass, and force in one frame $S$ to the corresponding values of these quantities in another frame $S'$, which moves with uniform velocity **v** with respect to $S$ along the common $x$-$x'$ axes.

Here we simply point out that these equations of transformation can be obtained in a perfectly straightforward way from the transformation equations already derived for the velocity components (Table 2-4). Once we have the force transformations, we can then use the Lorentz force law of classical electromagnetism (Eq. 3-9) to find how the numerical value of electric and magnetic fields depends on the frame of the observer. The details (see Ref. 6) are of no special interest to us here, but some of the conclusions reached are worth discussing in order to put special relativity theory in proper perspective.

We have seen in the last two chapters how kinematics and dynamics must be generalized from their classical form to meet the requirements of special relativity. And we saw earlier the role that optical experiments played in the development of relativity theory and the new interpretation that is given to such experiments. What remains, therefore, is to investigate classical electricity and magnetism in order to discover what modifications may need to be made there because of relativistic considerations. It turns out that Maxwell's equations are invariant under a Lorentz transformation and do not need to be modified (see Sec. 4.7 of Ref. 6 for proof). This result then completes the original program of finding the transformation (the Lorentz transformation) that keeps the velocity of light constant and finding the invariant form of the laws of mechanics and electromagnetism. The (Einstein) principle of relativity appears to apply to *all* the laws of physics.

Although relativity leaves Maxwell's equations of electromagnetism unaltered, it does give us a new point of view that enhances our understanding of electromagnetism. It is shown clearly in relativity that electric fields and magnetic fields have no separate meaning; that is, **E** and **B** do not exist independently as separate quantities but are interdependent. A field that is purely electric, or purely magnetic, in one inertial frame, for example, will have both electric and magnetic components in another inertial frame. One can find, from the force transformations, just how **E** and **B** transform from one frame to another. These equations of transformation are of much practical benefit, for we can solve difficult problems by choosing a reference system in which the answer is relatively easy to find and then transforming the results back to the system we deal with in the laboratory. The techniques of relativity, therefore, are often much simpler than the classical techniques for solving electromagnetic problems.

One striking result we obtain from relativity is this. If all we knew in electromagnetism was Coulomb's law, then, by using special relativity and the invariance of charge, we could prove that magnetic fields must exist. There is no need to postulate magnetic fields separately from electric fields. The magnetic field enters relativity in a most natural way as a field that is produced by a source charge in motion and that exerts a force on a test charge that depends on its velocity relative to the observer. Magnetism is simply a new word, a short-hand designation, for the velocity-dependent part of the force. In fact, starting only with Coulomb's law and the invariance of charge, we can derive (see Ref. 7) all of electromagnetism from relativity theory — the exact opposite of the historical development of these subjects.

## QUESTIONS

1. In view of the fact that particle speeds are limited by the speed of light, is "acceleration" a good word to use in relativity to describe the action of a force on a particle? Can you think of a more apt name? ("Ponderation"?)

2. Can we simply substitute $m$ for $m_0$ in classical equations to obtain the correct relativistic equations? Give examples.

3. Is it true that a particle that has kinetic energy must also have momentum? What if the particle has zero rest mass? Can a *system* of particles have kinetic energy but no momentum? Momentum but no kinetic energy?

4. A particle with zero rest mass (a neutrino, possibly) can transport momentum. How can this be in view of Eq. 3-5 $[p = m_0u/(1 - u^2/c^2)^{1/2}]$, in which we see that the momentum is directly proportional to the rest mass?

5. Distinguish between a variable-mass problem in classical physics and the relativistic variation of mass.

6. If a particle could be accelerated to a speed greater than the speed of light, what would be some of the consequences?

7. If zero-mass particles have a speed $c$ in one reference frame, can they be found at rest in any other frame? Can such particles have any speed other than $c$?

8. Does **F** equal $m$**a** in relativity? Does $m$**a** equal $d(m$**u**$)/dt$ in relativity?

**9.** What characteristic of a particle does the combination $\beta\gamma m_0 c$ represent? The symbols have their usual meanings.

**10.** We say that a l-keV electron is a "classical" particle, a l-MeV electron is a "relativistic" particle, and a l-GeV electron is an "extremely relativistic" particle. What exactly do these terms mean?

**11.** How many relativistic expressions can you think of in which the Lorentz factor $\gamma$ enters as a simple multiplier?

**12.** Discuss in detail, paying careful attention to signs, how the relation $u' = 2u$, discussed in connection with Fig. 3-1b, follows from the Galilean velocity transformation law (Eq. 2-19).

**13.** The total energy $E$ of the system is conserved in the collision shown in Fig. 3-2. However, the numerical value assigned to $E$ by the observer in frame $S$ is *not* the same as that assigned by the observer in frame $S'$. Is there a contradiction here?

**14.** Is a totally inelastic collision one in which all of the kinetic energy is lost, none remaining after the collision?

**15.** What determines whether a collision is elastic? Inelastic? Totally inelastic?

**16.** Here are some characteristics of particles or of groups of particles: rest mass, relativistic mass, total energy, kinetic energy, momentum. Which of these quantities is conserved in an elastic collision between two particles? In an inelastic collision?

**17.** In Section 3-5 we discussed the mode of operation of a proton synchrotron, designed to accelerate protons to very high energies. The precursor of this device was the conventional *cyclotron*, which resembled the synchrotron in broad outline except that neither the frequency of the accelerating oscillator nor the magnitude of the magnetic field changed with time. Discuss how this device might work well enough at low particle energies and point out the difficulties that would arise as the particle energy increased.

**18.** In a given magnetic field would a proton or an electron have the greater cyclotron frequency?

**19.** In the proton synchrotron at Fermilab (see Table 3-2), the radius of the magnet ring is 1000 m but the radius of the proton path is given as 750 m. How can these quantities be so different?

**20.** How can mass and energy be "equivalent" in view of the fact that they are totally different physical quantities, defined in different ways and measured in different units?

**21.** Is the rest mass of a stable composite particle (a uranium nucleus, say) greater or less than the sum of the rest masses of its constituents?

**22.** "The rest mass of the electron is 511 keV." What exactly does this statement mean?

**23.** "The relation $E = mc^2$ is essential to the operation of a power plant based on nuclear fission but has only a negligible relevance for a coal-fired plant." Is this a true statement?

24. A hydroelectric plant generates electricity because water falls under gravity through a turbine, thereby turning the shaft of a generator. According to the mass-energy concept, must the appearance of energy (the electricity) be identified with a (conventional) mass decrease somewhere? Where?

25. Exactly why is it that, pound for pound, nuclear explosions release so much more energy than do TNT explosions?

26. A hot metallic sphere cools off as it rests on the pan of a scale. If the scale were sensitive enough, would it indicate a change in rest mass?

27. A spring is kept compressed by tying its ends together tightly. It is then placed in acid and dissolves. What happens to its stored potential energy?

28. What role does potential energy play in the equivalence of mass and energy?

29. In Einstein's derivation of the $E = mc^2$ relation (see Example 12), how could he get away with using the classical (instead of the relativistic) expression for the aberration constant $\alpha$, a quantity that enters so centrally into his proof?

## PROBLEMS

$$c = 3.00 \times 10^8 \text{ m/s} = 300 \text{ km/ms} = 0.300 \text{ m/ns}$$
$$= 0.300 \text{ mm/ps}$$
$$c^2 = 8.99 \times 10^{16} \text{ J/kg} = 931 \text{ MeV/u}$$
$$1 \text{ u} = 1.66 \times 10^{-27} \text{ kg} \qquad 1 \text{ eV} = 1.60 \times 10^{-19} \text{ J}$$
$$1 \text{ GeV} = 10^3 \text{ MeV} = 10^6 \text{ keV} = 10^9 \text{ eV} = 1.60 \times 10^{-10} \text{ J}$$
$$\text{Electron rest mass} = 9.11 \times 10^{-31} \text{ kg} = 0.511 \text{ MeV}/c^2$$
$$\text{Proton rest mass} = 1.67 \times 10^{-27} \text{ kg} = 938 \text{ MeV}/c^2$$

1. **A flying brick.** Consider a building brick, of volume $V_0$ and rest mass $m_0$, at rest in a certain reference frame. Let the brick be viewed by a second observer for whom it is moving with speed $u$. (a) What is the volume of the brick from the point of view of this observer? (b) What mass will he measure for it? (c) What will be its density $\rho$, expressed in terms of the density $\rho_0$ of the resting brick? For what value of $u$ would the density measured by this observer increase by 1.0 percent over its rest value?

2. **Around the world in one second.** An electron is moving at a speed such that it could circumnavigate the earth at the equator in one second. (a) What is its speed, in terms of the speed of light? (b) Its kinetic energy $K$? (c) What percent error do you

make if you use the classical formula to calculate $K$?

3. **Doing work on an electron.** How much work must be done to increase the speed of an electron from rest (a) to $0.50c$? (b) To $0.990c$? (c) To $0.9990c$?

4. **True for electrons and baseballs.** What is the speed of a particle (a) Whose kinetic energy is equal to twice its rest energy, (b) Whose total energy is equal to twice its rest energy?

5. **A small change in speed but a large change in energy.** An electron has a speed of $0.9990c$. (a) What is its kinetic energy? (b) If its speed is increased by 0.05 percent, by what percent will its kinetic energy increase?

6. **The last steps are the hardest.** How much work must be done to increase the speed of an electron from (a) $0.180c$ to $0.190c$?

(b) 0.980c to 0.990c? Note that the speed increase (= 0.010c) is the same in each case.

★7. **The next step is always bigger.** An electron initially at rest in a certain reference frame is given a number of successive speed increases. Each increase takes it from its present speed 95 percent of the way to the speed of light. To what successive kinetic energy increases do the first six speed increases correspond?

8. **A useful high-energy approximation.** (a) Show that, for an extremely relativistic particle, the particle speed $u$ differs from the speed of light $c$ by

$$\Delta u = c - u = \left(\frac{c}{2}\right)\left(\frac{m_0 c^2}{E}\right)^2,$$

in which $E$ is the total energy. Find this quantity for an electron whose kinetic energy is (b) 100 MeV; (c) 25 GeV. (Electrons with energies in this latter range can be generated in the Stanford Linear Accelerator.)

9. **Classical physics and the speed of light.** (a) What potential difference would accelerate an electron to the speed of light, according to classical physics? (b) With this potential difference, what speed would the electron actually attain? (c) What would be its mass at that speed? (d) Its kinetic energy?

10. **Two ways to get it wrong.** The correct relativistic expression for kinetic energy is $K = (m - m_0)c^2$. (a) Show that the greatest possible percent error committed by using the classical expression $K = \frac{1}{2}m_0 v^2$ is 100 percent. (b) Show that the greatest possible percent error committed by using $K = \frac{1}{2}mv^2$ is 50 percent.

11. **Similar, but not the same.** Show that the relativistic expression for $K$ (see Eq. 3-11c) can be written as

$$K = \left(\frac{\gamma^2}{1 + \gamma}\right)m_0 u^2$$

$$\text{or} \quad K = \frac{p^2}{(1 + \gamma)m_0}.$$

Note that these expressions resemble in form the classical expressions:

$$K = \left(\frac{1}{2}\right)m_0 u^2 \quad \text{and} \quad K = \frac{p^2}{2m_0}.$$

Do the relativistic expressions reduce to the classical ones in the limit of low speeds?

12. **The lightest particles (?).** It is now thought that the neutrino, long believed to be massless, may have a small rest mass whose energy equivalent may be (let us assume) 25 eV. (a) What is the ratio of the rest mass of the electron to that of such a neutrino? (b) By how much would the speed of a 1.0-MeV neutrino with this rest energy differ from the speed of light? (Hint: See Problem 8.) (c) What kinetic energy must an electron have if its speed is to be the same as that of a 1.0-MeV neutrino?

13. **The fastest particles (?).** Cosmic rays, originating in deep space, fall steadily upon the earth. Most are protons, and a very small number of them have energies ranging up to $10^{20}$ eV. For such an extremely energetic proton, calculate (a) the difference between its speed and the speed of light and (b) its relativistic mass. (c) What would be the diameter of the earth's orbit about the sun as seen by an observer moving with such a speeding proton? In the rest frame of the solar system this diameter is $3.0 \times 10^8$ km. (See "Nature's Own Particle Accelerator," M. Mitchell Waldrop, *Sky and Telescope,* September, 1981.)

14. **Beta, gamma, and K (I).** Find the speed parameter $\beta$ and the Lorentz factor $\gamma$ for an electron whose kinetic energy is (a) 1.0 keV; (b) 1.0 MeV; (c) 1.0 GeV.

15. **Beta, gamma, and K (II).** Find the speed parameter $\beta$ and the Lorentz factor $\gamma$ for a particle whose kinetic energy is 10 MeV if the particle is (a) an electron; (b) a proton; (c) an alpha particle.

16. **A useful formula.** Show that the speed parameter $\beta$ and the Lorentz factor $\gamma$ are related by

$$\beta\gamma = \sqrt{\gamma^2 - 1}.$$

Show further that this relation remains valid at both the high-speed and the low-speed limits.

⋆**17. Pulsars as proton traps.** Pulsars are rapidly rotating neutron-stars. They are characterized by being small ($\sim$10 km diameter), by rotating rapidly, and by having an extremely intense associated magnetic field. This field is thought to "co-rotate" with the pulsar, with the field lines and the pulsar proper behaving like a rotating rigid body. Charged particles (which may be associated with the cosmic radiation) can get trapped along these field lines and co-rotate with the pulsar. The pulsar associated with the Crab nebula has a rotation period of 33 ms. (a) What is the greatest distance from its axis at which a trapped particle can co-rotate? (b) What is the kinetic energy of a co-rotating particle (assumed to be a proton) at 90 percent of this distance?

**18. Some missing algebra.** Derive Eq. 3-4a ($m = m_0/\sqrt{1 - u^2/c^2}$), starting from Eqs. 3-1, 3-2, and 3-3 and supplying all the missing steps.

**19. More missing algebra.** Supply all of the missing algebraic steps in the proof presented in Example 2.

**20. A singular value for the momentum.** A particle has a momentum equal to $m_0c$. (a) What is its speed? (b) Its mass? (c) Its kinetic energy?

**21. A memory-jogging triangle.** (a) Suppose that the mnemonic triangle of Fig. 3-3 represents a proton. By measurement, what is the kinetic energy of the proton? (b) What is the kinetic energy if the particle is an electron? (c) Draw triangles to represent a 10-keV electron and a 2.5-MeV electron.

**22. Three useful relationships.** (a) Verify Eq. 3-13a connecting $K$ and $p$ by eliminating $u$ between Eqs. 3-5 and 3-11b. (b) Derive the following useful relations among $p$, $E$, $K$, and $m_0$ for relativistic particles, starting from Eqs. 3-13a and 3-13b:

$$(1)\ K = c\sqrt{m_0^2c^2 + p^2} - m_0c^2.$$

$$(2)\ p = \frac{\sqrt{K^2 + 2m_0c^2K}}{c}.$$

$$(3)\ m_0 = \frac{\sqrt{E^2 - p^2c^2}}{c^2}.$$

(c) Show that these relationships remain valid as $u \to 0$.

**23. Energy and momentum—a graphical study.** Plot the total energy $E$ against the momentum $p$ for a particle of rest mass $m_0$ under three assumptions: (a) that the kinetic energy of the particle is given by the classical expression $p^2/2m_0$; (b) that the particle is a relativistic particle; and (c) that the particle is a relativistic particle but has zero rest mass. (d) In what region does curve (b) approach curve (a), and in what region does curve (b) approach curve (c)? (e) Explain briefly the physical significance of the intercepts of the curves with the axes and of their slopes (derivatives). (In classical physics, the energy of a single particle is defined only to within an arbitrary constant. Assume, in (a), that this arbitrary constant is the same as that fixed by relativity, namely, that the energy of a particle at rest, $E_0$, equals $m_0c^2$.)

**24. Three particles compared.** Consider the following, all moving in free space: a 2.0-eV photon, a 0.40-MeV electron and a 10-MeV proton. (a) Which is moving the fastest? (b) The slowest? (c) Which has the greatest momentum? (d) The least? (*Note:* A photon is a light-particle of zero rest mass.)

**25. Conditions at zero rest mass.** (a) Show that a particle that travels at the speed of light must have zero rest mass. (b) Would such a particle—such as light—travel with speed $c$ in all inertial reference frames? (c) Show that for a particle of zero rest mass, $u = c$, $\gamma = 1$, $K = E$, and $p = K/c$.

26. **A practical unit for momentum.** High-energy particle physicists commonly use the GeV ( $= 10^9$ eV $= 1.60 \times 10^{-10}$ J) as a practical laboratory unit in which to measure the kinetic energy of a particle. (*a*) Show that a similarly practical unit for the momentum of a particle is the GeV/*c* and that 1 GeV/*c* $= 5.33 \times 10^{-19}$ kg · m/s. What is the momentum of a proton whose kinetic energy *K* is (*b*) 5.0 GeV? (*c*) 500 GeV?

27. **Momentum and energy for high-speed particles.** In particle physics the momenta of energetic particles are usually reported in units such as GeV/*c*. One reason for this is that, as $u \to c$, the relationship $p = E/c$ (which is strictly true only for particles with zero rest mass) becomes more and more correct. In the extreme relativistic realm, then, one number gives both the energy and the momentum of the particle. Verify these considerations by filling in the following table, which refers to a proton.

| $\beta$ ( $= u/c$) | 0.80 | 0.90 | 0.99 | 0.999 | 0.9999 |
|---|---|---|---|---|---|
| *E*,  GeV | | | | | |
| *p*,  GeV/*c* | | | | | |

28. **Two fast particles compared.** A particle has a speed of $0.990c$ in a laboratory reference frame. What are its kinetic energy, its total energy, and its momentum if the particle is (*a*) a proton or (*b*) an electron?

29. **Finding the rest mass.** (*a*) If the kinetic energy *K* and the momentum *p* of a particle can be measured, it should be possible to find its rest mass $m_0$ and thus identify the particle. Show that

$$m_0 = \frac{(pc)^2 - K^2}{2Kc^2}.$$

(*b*) Show that this expression reduces to an expected result as $u/c \to 0$, in which *u* is the speed of the particle. (*c*) Find the rest mass

of a particle whose kinetic energy is 55.0 MeV and whose momentum is 121 MeV/*c*; express your answer in terms of the rest mass of the electron.

30. **Decay in flight.** The average lifetime of muons at rest is 2.20 μs. A laboratory measurement on the decay in flight of the muons in a beam emerging from a particle accelerator yields an average lifetime of 6.90 μs. (*a*) What is the speed of these muons in the laboratory? (*b*) The relativistic mass (in terms of $m_e$, the rest mass of the electron)? (*c*) The kinetic energy? (*d*) The momentum? The rest mass of a muon is 207 times greater than that of an electron.

31. **Action in the upper atmosphere.** In a high-energy collision of a primary cosmic-ray particle near the top of the earth's atmosphere, 120 km above sea level, a pion is created with a total energy *E* of $1.35 \times 10^5$ MeV, traveling vertically downward. In its proper frame this pion decays 35 ns after its creation. At what altitude above sea level does the decay occur? The rest energy of a pion is 139.6 MeV.

★32. **A center-of-mass reference frame.** Proton *A*, whose speed is $0.990c$ in laboratory reference frame *S*, is approaching proton *B*, which is at rest in that frame. The center of mass of these two protons (a point) moves in the laboratory with a certain constant velocity. Consider a second reference frame $S'$ moving in the laboratory with this same velocity. Calculate the kinetic energy *K*, the total energy *E*, and the momentum *p* of each proton in each reference frame, recording your answers in the table below.

| Proton | Frame | *K* | *E* | *p* |
|---|---|---|---|---|
| *A* | *S* | | | |
| *B* | *S* | | | |
| *A* | $S'$ | | | |
| *B* | $S'$ | | | |

33. **Solving the equation of motion.**   Equation 3-17 shows Newton's second law of motion extended to relativistic form. (*a*) For the special case in which the force **F** acting on the particle and its velocity **u** point in the same direction, show that this equation reduces to

$$\left(1 - \frac{u^2}{c^2}\right)^{-3/2} du = \frac{F}{m_0} dt,$$

in which we assume that *F* does not depend on *u*.

(*b*) Show that the solution to this differential equation is

$$u = \frac{(F/m_0)t}{\sqrt{1 + (F/m_0 c)^2 t^2}}.$$

(*c*) Show that this solution can also be written as

$$t = \frac{(m_0/F)u}{\sqrt{1 - u^2/c^2}},$$

in which we have required that $u = 0$ when $t = 0$. (*d*) Do these last two equations reduce to expected results for $u \rightarrow 0$? For $t \rightarrow \infty$?

34. **The force law in relativity—two special cases.**   In general (see Eq. 3-17), the acceleration **a** of a particle is not parallel to the force **F** acting on it. (*a*) Show that there are two simple special cases in which the acceleration *is* parallel to the force, namely, when **F** is parallel to the velocity **u** and when **F** is perpendicular to **u**. (*b*) Give a physical example of each such special case. (*c*) In view of the fact that we can always resolve a force into two such components (one parallel to **u** and one perpendicular to it), why is **F** not *always* parallel to **a**?

✶35. **Accelerated motion in a straight line.**   In Example 6, assume that the accelerating potential (= 4.50 MV) is applied (with a uniform gradient) between the ends of an accelerating tube 3.00 m long. In that example the speed of the electron upon emerging from the accelerator was shown to be $0.9948c$. What was the duration of the acceleration? (*Hint:* Use the result of Problem 33).

✶36. **A familiar collision analyzed.**   In the collision of Fig. 3-2, assume that $m_0 = 1.000$ u and $u = 0.200c$. (*a*) Calculate $M_0$ and $u'$. (*b*) What is the change in kinetic energy during the collision from the point of view of frame *S*? (*c*) From the point of view of frame *S'*? (*d*) What is the change in rest energy during the collision?

37. **A simple inelastic collision.**   Two identical objects, each of rest mass $m_0$, moving with equal but opposite velocities of $0.60c$ in the laboratory reference frame, collide and stick together. The resulting particle has a rest mass $M_0$. Express $M_0$ in terms of $m_0$.

38. **Another simple inelastic collision.**   A body of rest mass $m_0$, traveling initially at a speed of $0.60c$, makes a completely inelastic collision with an identical body that is initially at rest. (*a*) What is the rest mass of the resulting single body? (*b*) What is its speed?

39. **The previous collision generalized.**   Let the initial speed of the particle of rest mass $m_0$ in Problem 38 be generalized to $\beta c$, all other conditions of that problem remaining unchanged. (*a*) Show that the rest mass $M_0$ of the resulting single particle is given by

$$M_0 = \sqrt{2(\gamma + 1)} m_0.$$

(*b*) Show that the speed *U* of that particle is given by

$$U = \sqrt{\frac{\gamma - 1}{\gamma + 1}} c.$$

(*c*) Verify that, for $\beta = 0.60$, these formulas reproduce the numerical answers given in Problem 38.

40. **Not the simplest reference frame!**   Two high-speed particles, each having rest mass $m_0$, approach each other on course for a head-on collision. One has a speed of $0.80c$ and the other a speed of $0.60c$. Assuming that the resulting collision is *completely inelastic*, answer the following questions in terms of $m_0$ and *c*. (*a*) What is the momentum after the collision? (*b*) What value does Newtonian mechanics predict for this quan-

tity? (c) What is the total energy after the collision? (d) What is the total rest mass after the collision? (e) What is the total kinetic energy after the collision?

41. **The completely inelastic collision.** Verify that total energy $E$ is conserved, in each reference frame, for the completely inelastic collision of Fig. 3-2.

42. **An elastic collision.** Consider the following head-on elastic collision. Particle 1 has rest mass $2m_0$ and particle 2 has rest mass $m_0$. Before the collision, particle 1 moves toward particle 2, which is initially at rest, with speed $u$ ($= 0.600c$). After the collision each particle moves in the forward direction with speeds of $u_1$ and $u_2$, respectively. (a) Apply the laws of conservation of total energy (or, equivalently, of relativistic mass) and of relativistic momentum to this collision and solve the resulting equations to find $u_1$ and $u_2$. (b) Calculate the initial and the final kinetic energy and show that kinetic energy is indeed conserved in this elastic collision. (*Hint:* The relationship $\beta\gamma = \sqrt{\gamma^2 - 1}$ may be found useful.)

★43. **The great neutron shoot-out.** Sue and Jim are two experimenters at rest with respect to one another at different points in space. They "fire" neutrons at each other, each neutron leaving its "gun" with a relative speed of $0.60c$. Jim makes five observations about what is going on: (a) "My separation from Sue is 10 km." (b) "The speed of 'my' neutrons is $0.60c$." (c) "Two of our neutrons have collided; relativistic momentum and kinetic energy are conserved." (d) "After this collision, one of the neutrons was scattered through an angle of 30°." (e) "I am firing neutrons at the rate of 10,000 $s^{-1}$." For each of these observations, state the corresponding observation that would be reported by a third person (Sam, say), who is in a frame $S$ chosen so that Sue's neutrons are at rest in it.

44. **The relativistic rocket.** A rocket with initial mass $m_i$ (fuel + payload) is accelerated from rest to a final speed $v$ that is comparable to the speed of light. It can be shown that the mass of the rocket $M_p$ (payload alone) when the speed $v$ has been achieved is given by

$$\frac{M_i}{M_p} = \left(\frac{1 + \beta}{1 - \beta}\right)^{c/2u},$$

in which $u$ is the speed of the exhaust relative to the rocket and $\beta = v/c$. Hence, to keep the ratio $M_i/M_p$ small, large exhaust speeds are needed. (a) Using nuclear fusion to generate the exhaust, exhaust speeds of $c/7$ are theoretically possible. Suppose that, for interstellar travel, a final cruising speed $v$ of $0.99c$ is desired. Calculate $M_i/M_p$. (b) For a round trip to a star—Sirius, say—the rocket must be accelerated from rest near the earth to cruising speed $v$ ($= 0.99c$), brought to a stop at Sirius, accelerated from rest for the return trip when the visit is over, and finally brought to rest at home base. Calculate $M_i/M_p$ for this round trip, where $M_p$ is now the mass of the rocket when the round trip is over. (c) The greatest possible exhaust speed ($u = c$) occurs if the exhaust is electromagnetic radiation, generated perhaps by matter—antimatter annihilation. Recalculate the ratios in (a) and (b) for such a rocket. (d) In view of all of the above, what do you think of the possibility of interstellar travel?

45. **Fast particles curve less than slow ones.** What is the radius of curvature of an electron moving at right angles to a uniform magnetic field of 0.10 T at a speed of (a) $0.9c$? (b) $0.99c$? (c) $0.999c$? (d) $0.9999c$?

46. **Identifying particles.** Ionization measurements show that a particular nuclear particle carries a double charge ($= 2e$) and is moving with a speed of $0.710c$. Its measured radius of curvature in a magnetic field of 1.00 T is 6.28 m. Find the rest mass of the particle and identify it. (*Hint:* Light nuclear particles are made up of neutrons [which carry no charge] and protons [charge $= +e$], in roughly equal numbers. Take the rest mass of either of these particles to be 1.00 u.)

47. **A curving relativistic electron.** A 2.50-MeV electron moves at right angles to a magnetic field in a path whose radius of curvature is 3.0 cm. (*a*) What is the magnetic field *B*? (*b*) By what factor does the relativistic mass of the electron exceed its rest mass?

48. **A curving cosmic-ray proton.** A 10-GeV proton in the cosmic radiation approaches the earth in the plane of its geomagnetic equator, in a region over which the earth's average magnetic field is $5.5 \times 10^{-5}$ T. What is the radius of its curved path in that region?

49. **Cosmic rays in the sun's magnetic field.** A cosmic-ray proton of kinetic energy 10 GeV may experience an effective magnetic field due to the sun of $\sim 2 \times 10^{-10}$ T. What will be the radius of curvature of the path of such a proton? Compare this radius to the radius of the earth's orbit about the sun ($1.5 \times 10^{11}$ m).

50. **Cosmic-ray protons in the galactic magnetic field.** Cosmic ray particles (presumably protons) with energies as large as $10^{20}$ eV have been detected as they enter our atmosphere. What is the radius of curvature of such an extremely energetic proton as it moves in the extremely weak magnetic field ($10^{-9} - 10^{-10}$ T) that permeates our galaxy? Compare your answer with the radius of (the luminous portion of) the galaxy, which is about 80,000 ly. Assume for simplicity that the galactic magnetic field is uniform and that the proton moves at right angles to it.

51. **The ultimate terrestrial accelerator (I).** Imagine a proton synchrotron extending around the earth at its equator, with a maximum magnetic field of 5.0 T. (*a*) What would be the energy of a proton circulating in such a path? (*b*) What would be the ratio of its relativistic mass to its rest mass?

52. **The ultimate terrestrial accelerator (II).** The linear electron accelerator at Stan-

ford is about two miles long and has an acceleration gradient of about 15 MeV/m. Imagine a similar accelerator built into a tunnel through the earth, along a diameter. (*a*) What electron energy would be achieved by such a device, assuming that the same acceleration gradient could be maintained? (*b*) What would be the ratio of the relativistic mass to the rest mass for such electrons?

53. **The binding energy of carbon-12.** The nucleus $^{12}$C consists of six protons ($^1H$) and six neutrons (*n*) held in close association by strong nuclear forces. The rest masses are

$$
\begin{array}{ll}
^{12}\text{C} & 12.000000 \text{ u,} \\
^{1}\text{H} & 1.007825 \text{ u,} \\
n & 1.008665 \text{ u.}
\end{array}
$$

How much energy would be required to separate a $^{12}$C nucleus into its constituent neutrons and protons? This energy is called the *binding energy* of the $^{12}$C nucleus. (*Note:* The masses given are really those of neutral atoms, but the extranuclear electrons have relatively negligible binding energy and are of equal number both before and after the breakup of the $^{12}$C nucleus.)

54. **A neutron-capture process.** A helium-3 nucleus (nuclear mass = 3.01493 u) captures a slow neutron (mass = 1.00867 u) to form a nucleus of helium-4 (nuclear mass = 4.00151 u), according to the scheme

$$^{3}\text{He} + n \rightarrow {}^{4}\text{He} + \gamma.$$

The symbol $\gamma$ represents an emitted gamma ray. What is its energy?

55. **Spontaneous decay.** A body of mass *m* at rest breaks up spontaneously into two parts with rest masses $m_1$ and $m_2$ and respective speeds $u_1$ and $u_2$. Show that $m > m_1 + m_2$.

★56. **The decay of a pion.** A charged pion at rest in the laboratory decays according to the scheme

$$\pi \rightarrow \mu + \nu,$$

in which $\mu$ represents a muon (rest energy = 105.7 MeV) and $\nu$ a neutrino (rest energy taken as zero). If the measured kinetic energy of the muon is 4.1 MeV, what rest energy may be calculated for the pion?

★ **57. Fixed-target collisions.**　(*a*) A proton accelerated in a proton synchrotron to a kinetic energy $K$ strikes a second (target) proton at rest in the laboratory. The collision is entirely inelastic in that the rest energy of the two protons, plus all of the kinetic energy consistent with the law of conservation of momentum, is available to generate new particles and to endow them with kinetic energy. Show that the energy available for this purpose is given by

$$\varepsilon = 2m_0c^2 \sqrt{1 + \left(\frac{K}{2m_0c^2}\right)}.$$

(*b*)　How much energy is made available with 100-GeV protons are used in this fashion? (*c*) What proton energy would be required to make 100 GeV available? (*Note:* Compare Problem 58.)

**58. Center-of-mass collisions.**　(*a*) In modern experimental high-energy physics, energetic particles are made to circulate in opposite directions in so-called storage rings and permitted to collide head-on. In this arrangement each particle has the same kinetic energy $K$ in the laboratory. The collisions may be viewed as totally inelastic, in that the rest energy of the two colliding protons, plus all available kinetic energy, can be used to generate new particles and to endow them with kinetic energy. Show that the available energy in this arrangement can be written in the form (compare Problem 57)

$$\varepsilon = 2m_0c^2 \left(1 + \frac{K}{m_0c^2}\right).$$

(*b*) How much energy is made available when 100-GeV protons are used in this fashion? (*c*) What proton energy would be required to make 100 GeV available? (*Note:* Compare your answers with those in Problem 57, which describes another—less energy-effective—bombarding arrangement.)

**59. Nuclear recoil during gamma emission.**　The nucleus of a carbon atom initially at rest in the laboratory is in a so-called excited, or unstable, state. It goes to its stable ground state by emitting a pulse of radiation (gamma ray) whose energy is 4.43 MeV and simultaneously recoiling. Thus

$$^{12}\text{C}^* \rightarrow {}^{12}\text{C} + \gamma,$$

in which $\gamma$ represents the gamma ray and the asterisk signifies an excited state. The atom in its final state has a rest mass of 12.0 u. (*a*) What is the momentum of the recoiling carbon atom, as measured in the laboratory? (*b*) What is the kinetic energy of the recoiling atom? (*Hint:* The speed of the relatively massive recoiling carbon atom will be so small that the classical expressions for its kinetic energy and momentum can be used without appreciable error.)

**60. Atomic recoil during photon emission.**　An excited atom, with excitation energy $E$ and rest mass $m_0$, is initially at rest. It releases this energy by emitting a pulse of radiation (photon) of energy $E'$ and simultaneously recoiling. Because of the kinetic energy given to the recoiling atom, the energy $E'$ of the photon will be (slightly) less than $E$. Show that, to a close approximation,

$$E' = E\left(1 - \frac{E}{2m_0c^2}\right).$$

(*Hint:* The speed of the relatively massive recoiling atom will be low enough that the classical expressions for the kinetic energy and the momentum of the atom may be used without appreciable error. The answer given involves an approximation based on the binomial expansion.)

**61. Deriving a conversion factor.**　(*a*) Prove that 1 u = 931 MeV/$c^2$. (*b*) Find the energy equivalent to the rest mass of an electron and (*c*) of a proton.

**62. A real fast ball!**　A baseball has a mass of 140 g. (*a*) What energy must be supplied to it if its relativistic mass is to exceed its rest mass by 0.1 percent? (*b*) For how long a time would the full output of a 1000-MW power plant be required to supply this energy?

**63. A low-speed space probe.** *Voyager 2*, an 810-kg planetary probe, flew by Saturn in 1981, achieving speeds in excess of 54,000 mi/h. At this speed, what mass increment is associated with the kinetic energy of the probe?

**64. A good student project (?).** A metric ton of water (= 1000 kg) is heated from the freezing point to the boiling point. By how much does its mass increase?

**65. Lots of energy in a penny.** (*a*) What is the equivalent in energy units of the mass of a penny (= 3.1 g)? (*b*) How long would a 1000-MW power plant have to run to generate this amount of energy?

**66. Water is heavier than ice (but not by much).** Find the fractional increase in mass when an ice cube melts. The energy required is $3.35 \times 10^5$ J/kg.

**67. About the weight of 30 gal of gasoline.** The United States consumed about $2.2 \times 10^{12}$ kWh of electrical energy in 1979. How much matter would have to vanish to account for the generation of this energy? Does it make any difference to your answer if this energy is generated in oil-burning, nuclear, or hydroelectric plants?

**68. Drop a tablet in your tank!** A 5-grain aspirin tablet has a mass of 320 mg. For how many miles would the energy equivalent of this mass, in the form of gasoline, power an automobile? Assume 30 mi/gal and a heat of combustion of $1.3 \times 10^8$ J/gal for the gasoline.

**69. A ton of sunlight.** The sun radiates energy at the rate of $4.0 \times 10^{26}$ W. How many "tons of sunlight" (mass equivalent) does the earth intercept each day?

**70. An enormous source of energy.** Quasars are thought to be the nuclei of active galaxies in the early stages of their formation. A typical quasar radiates energy at the rate of $10^{41}$ W. At what rate is the mass of this quasar being reduced to supply this energy? Express your answer in solar mass units per year, where one solar mass unit, (smu = $2 \times 10^{30}$ kg), is the mass of our sun.

**71. Nuclear and TNT explosions compared.** (*a*) How much energy is released in the explosion of a fission bomb containing 3.0 kg of fissionable material? Assume that 0.10 percent of the rest mass is converted to released energy. (*b*) What mass of TNT would have to explode to provide the same energy release? Assume that each mole of TNT liberates 3.4 MJ of energy on exploding. The molecular weight of TNT is 0.227 kg/mol. (*c*) For the same mass of explosive, how much more effective are nuclear explosions than TNT explosions? That is, compare the fractions of the rest mass that are converted to energy in each case.

**72C. Using your calculator (I).** Store in your hand-held programmable calculator the rest energies of the electron, the muon, and the proton (see Appendix), and also make provision to enter the rest energy of any other particle that you may choose. Write a program that will allow you to select one of these particles and that will accept as input data any *one* of the following five properties of that particle: its kinetic energy, its total energy, its momentum, its Lorentz factor $\gamma$, or its speed parameter $\beta$. The output is to be the remaining four quantities. (*Note:* Your calculator may overflow if you try to calculate $\beta$ for particles— 20-GeV electrons, say—whose speeds are very close to the speed of light. See Problem 74C.)

**73C. Checking it out (I).** Check out the program you wrote in Problem 72C by answering these questions. (*a*) What is the momentum of an electron whose kinetic energy is 2.00 eV? 2.00 keV? 2.00 MeV? 2.00 GeV? (b) For what kinetic energies will an electron and a proton each have a momentum of 5.00 MeV/c? (c) An electron, a muon, and a proton each have a kinetic energy of 10.0 MeV. What are their speed parameters? (d) An electron, a muon, and a proton each have a relativistic mass that is three times their rest mass. What are their kinetic energies?

**74C. Using your calculator (II).** For particles whose speed parameters are sufficiently close to unity, the program called for in Problem 72C will yield a meaningless result for β because of calculator overflow. In such cases it is useful to calculate $1 - β$, using an approximate formula appropriate to the extreme relativistic case. To study the transition from the relativistic to the extreme relativistic case, write a program for your hand-held calculator that will accept as an input the kinetic energy of an electron and will display as successive outputs: (1) the speed parameter β, calcu-

lated exactly; (2) the quantity $1 - β$, also calculated exactly; (3) the quantity $1 - β$, calculated using an approximate formula appropriate to the extreme relativistic case; and (4) the percent difference between these last two quantities. (*Hint:* Store the rest energy of the electron, which is 0.511003 MeV, in your calculator. For the extreme relativistic formula, use $(1 - β) = (1/2)(m_0 c^2 / K)^2$, in which $K$ is the kinetic energy and $m_0 c^2$ is the rest energy.)

**75C. Checking it out (II).** (a) Run the program that you have written in Problem 74 for electron kinetic energies extending from a few keV to several hundred GeV and get a feeling for the transition from the relativistic to the extremely relativistic case. (b) At what electron energy do the exact and the approximate formulas for $1 - β$ differ by 10 percent? By 1.0 percent? (c) At about what electron kinetic energy do the predictions of the exact formula for β break down totally because of calculator overflow?

# REFERENCES

1. R. Kollath and D. Menzel, "Measurement of the Charge on Moving Electrons," *Z. Phys.* **134,** 530 (1953).
2. W. Bertozzi, "Speed and Kinetic Energy of Relativistic Electrons," *Am. J. Phys.,* **32,** 551 (1964).
3. R. R. Wilson, "The Batavia Accelerator," *Sci. Am.* (February 1974); R. R. Wilson, "The Tevatron," *Phys. Today* (October 1977).
4. R. T. Weidner, "On Weighing Photons," *Am. J. Phys.,* **35,** 443 (1967).
5. See *The Principle of Relativity* (Dover, New York, 1953), p. 29. This book is a collection of English translations of original papers by Einstein, Lorentz, and others.
6. Robert Resnick, *Introduction to Special Relativity* (Wiley, New York, 1968).
7. Edward M. Purcell, *Electricity and Magnetism* (McGraw-Hill, New York, 1965).
8. Mark P. Haugan and Clifford M. Will, "Modern Tests of Special Relativity," *Physics Today* (May 1987).

# The Geometric Representation of Spacetime

*Oh, that Einstein, always cutting lectures—I really would not have believed him capable of it.*

*Hermann Minkowski* (ca. 1908)

## A-1 Spacetime Diagrams

We have seen that in classical physics it is proper to treat the space and time coordinates separately. In relativity, however, it is natural to treat them together, their intimate interconnection being clearly displayed in the Lorentz transformation equations; see Tables 2-2 and 2-3. The common use of the single word "spacetime" (without a hyphen) to represent the coordinate description of events is symbolic of the general acceptance of this view.

As we have learned, it was Einstein [1] who first set forth, in his special theory of relativity, the physical basis for the proper description of events in space and time. Shortly afterwards the mathematician Hermann Minkowski (who, incidentally, had formerly been Einstein's mathematics professor in Zurich) [2] presented a simple and symmetrical geometric representation of these ideas, a representation that permits a ready understanding in geometric terms of such matters as the relativity of simultaneity, the length contraction, and the time dilation, including their reciprocal nature.

In what follows, we shall consider only one space axis, the $x$ axis, and shall ignore the $y$ and $z$ axes. We lose no generality by this algebraic simplification, and this procedure will enable us to focus more clearly on the interdependence of space and time and its geometric representation. The coordinates of an event are given, then, by $x$ and $t$. All possible spacetime coordinates can be represented on a spacetime diagram in which the space axis is horizontal and the time axis is vertical. It is convenient to keep the dimensions of the coordinates the same; this is easily done by multiplying the time $t$ by the universal constant $c$, the velocity of light. Let $ct$ be represented by the symbol $w$. Then, the Lorentz transformation equations (see Table 2-2 and Problem 11 of Chapter 2) can be written as follows:

$$
\begin{array}{llll}
(a) & x' = \gamma(x - \beta w) & \qquad (a') & x = \gamma(x' + \beta w') \\
(b) & w' = \gamma(w - \beta x) & \qquad (b') & w = \gamma(w' + \beta x')
\end{array}
\qquad \text{(A-1)}
$$

Notice the symmetry of this form of the equations.

To represent the situation geometrically, we begin by drawing the $x$ and $w$ axes of frame $S$ at right angles to one another, as in Fig. A-1. If we want to represent a moving particle in this frame, we draw a curve, called the *world line* of the particle, which gives the loci of spacetime points corresponding to the motion. The tangent to the world line at any point makes an angle $\theta$ with the direction of the time axis that is given by $\tan \theta = dx/dw = (dx/dt)(1/c) = u/c$. Because we must have $u < c$ for a material particle, the angle $\theta$ at any point on its world line must always be less than $45°$. If the particle is at rest, say, at position $x_0$ on the $x$ axis of Fig. A-1, its world line is parallel to the $w$ axis, with $\theta \ (= \tan^{-1} u/c) = 0$ at all points. For a light ray traveling along the $x$ axis we have $u = c$, so its world line is a straight line making an angle of $45°$ with the axes.

Consider now the primed frame ($S'$), which moves relative to $S$ with a velocity **v** along the common $x$-$x'$ axis. The equation of motion of the origin of $S'$ relative to $S$ can be obtained by setting $x' = 0$; from Eq. A-1a, we see that this corresponds to $x = \beta w$. We draw the line $x' = 0$ (that is, $x = \beta w$) on our diagram (Fig. A-2) and note that since $v < c$ and $\beta < 1$, the angle this line makes with the $w$ axis, $\phi \ (= \tan^{-1} \beta)$, is less than $45°$. Just as the $w$ axis corresponds to $x = 0$ and is the time axis in frame $S$, so the line $x' = 0$ gives the time axis $w'$ in $S'$. Now, if we draw the line $w' = 0$ (giving the location of clocks that read $t' = 0$ in $S'$), we shall have the space axis $x'$. That is, just as the $x$ axis corresponds to $w = 0$, so the $x'$ axis corresponds to $w' = 0$. But, from Eq. A-1b, $w' = 0$ gives us $w = \beta x$ as the equation of this axis on our $w$-$x$ diagram (Fig. A-2). The angle between the space axes is the same as that between the time axes. Note that, for simplicity, we have shown in Fig. A-2 only the quadrant in which both $x$ and $w$ are positive.

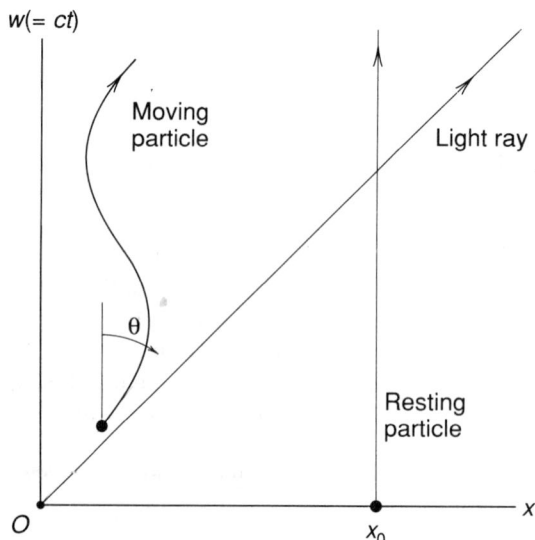

**FIGURE A-1.** The *world lines* of light and some particles.

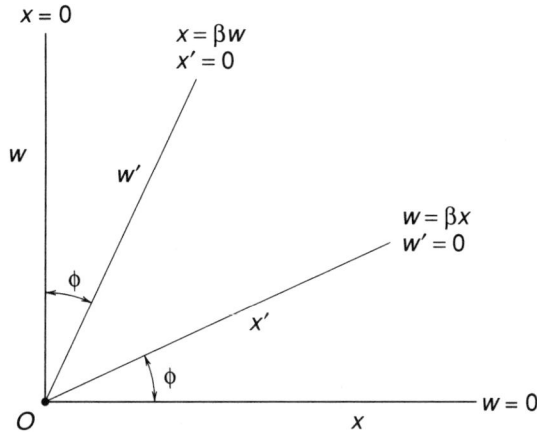

**FIGURE A-2.**  The *Minkowski diagram* for frames *S* and *S'*.

You should compare Fig. A-2 carefully with the standard representation of Fig. 1-1, which we have used exclusively in the main body of the text. A point in the coordinate reference frames of Fig. 1-1 shows only the space coordinates of the event to which it corresponds; the time of occurrence of the event must be given separately. A point on the Minkowski diagram of Fig. A-2, however, shows both the space and the time coordinates of the event in a single geometric representation.

## A-2  Calibrating the Spacetime Axes

Before we can make practical use of the spacetime diagram we must establish scales on its $x$, $w$ and its $x'$, $w'$ axes. We can use the Lorentz transformation equations of Eq. A-1 for this purpose. Consider first point $O$, located at the common origin of the two pairs of axes in Fig. A-3. It has coordinates $x = w = 0$ and $x' = w' = 0$, and the event to which it corresponds is the coincidence in time of the origins of the $S$ and $S'$ reference frames.

Point $P_1$ on the $x'$ axis of Fig. A-3 has been chosen as a point to which we wish to assign the value $x' = 1$, representing a unit of length on this axis. As for all points on the $x'$ axis, the time coordinate $w'$ of $P_1$ is zero. Putting $x' = 1$ and $w' = 0$ into Eq. A-1a′ yields, by simple inspection, $x = \gamma$ for the $x$ coordinate of $P_1$. With this information we can easily construct numerical scales for both the $x$ and the $x'$ axes, based on our initially assumed unit length.

Consider now point $P_2$ on the $w'$ axis of Fig. A-3, to which we wish to assign the value $w' = 1$, representing a unit of time (measured in terms of $ct'$, to be sure) on that axis. We wish the scales on both the $x'$ and the $w'$ axes to be based on the same unit length, so we choose to locate $P_2$ so that the line segment $OP_2$ is equal in length to the segment $OP_1$. As for all points on the $w'$ axis, the space coordinate $x'$ of $P_2$ is zero. Putting $w' = 1$ and $x' = 0$ into Eq. A-1b′ yields, again by simple inspection, $w = \gamma$ for the $w$ coordinate of $P_2$. We are now able to construct

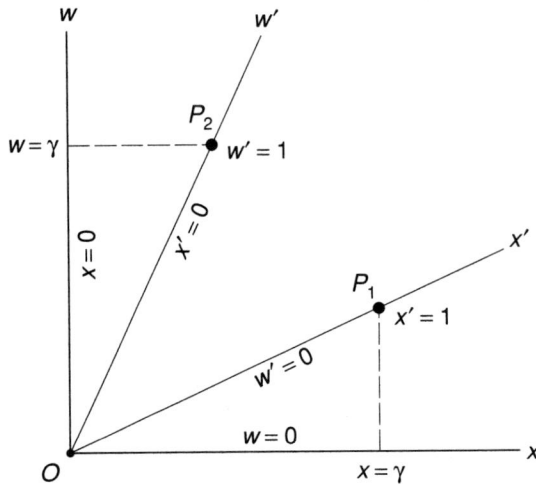

**FIGURE A-3.** Establishing the scales on the spacetime axes.

numerical scales for both the $w$ and the $w'$ axes, based on the same unit length as we assumed in calibrating the space axes.

To gain some physical familiarity with the Minkowski diagram, let us consider a clock at rest at the origin of the $S'$ frame. For that clock we have $x' = 0$ (always), so events involving it must correspond to points along the $w'$ axis of Fig. A-3. Point $O$, which is on that line, could represent the coincidence of the clock hand with a fiducial marker on the clock face, corresponding to zero time. Point $P_2$, whose time coordinate in the $S'$ frame gives unit time ($w' = 1$) on that resting clock, is also on that line. The event represented by $P_2$ might correspond to a second coincidence of the clock hand with the fiducial marker. In frame $S$, however, the clock would be seen as a moving clock. We have seen above that $w' = 1$ in the $S'$ frame corresponds to $w = \gamma$ in the $S$ frame. Thus, by $S$-frame clocks, the unit time interval of the $S'$ clock would be recorded as $\gamma$, corresponding exactly to the time dilation effect described by Eq. 2-14$b$.

In Fig. A-4 we show the calibration of the axes of the frames $S$ and $S'$, the unit time interval along $w'$ being a longer line segment than the unit time interval along $w$ and the unit length interval along $x'$ being a longer line segment than the unit length interval along $x$. The first thing we must be able to do is to determine the spacetime coordinates of an event such as $P$ directly from the Minkowski diagram. To find the space coordinate of the event, we simply draw a line parallel to the time axis from $P$ to the space axis. The time coordinate is given similarly by a line parallel to the space axis from $P$ to the time axis. The rules hold equally well for the primed frame as for the unprimed frame. In Fig. A-4, for example, the event $P$ has the spacetime coordinates $x = 3.0$ and $w = 2.5$ in $S$ (long dashed lines) and spacetime coordinates $x' = 2.0$ and $w' = 1.2$ in $S'$ (short dashed lines). Figure A-4 was drawn assuming that $\beta = 0.50$, which yields $\gamma = 1.15$. Using these values for $\beta$ and $\gamma$, you can readily derive the $S$-frame coordinates from the $S'$-frame

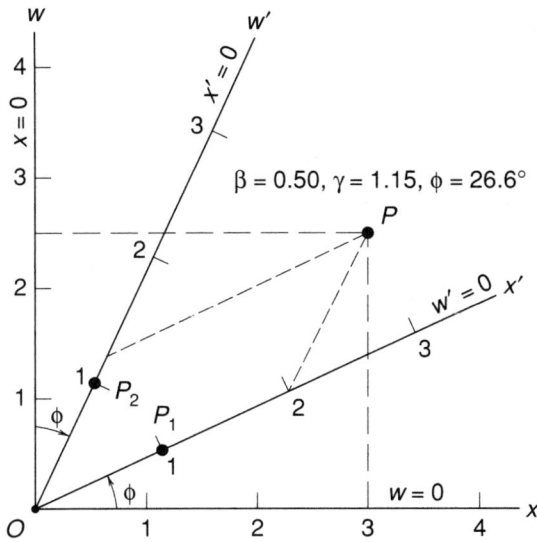

**FIGURE A-4.**  Calibrating the axes of the frames S and S'.

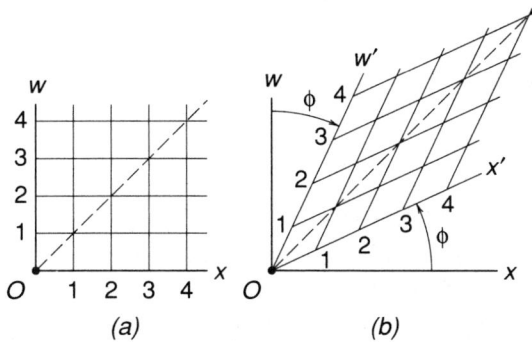

**FIGURE A-5.**  An orthogonal reference frame, (a), transforms into a nonorthogonal one, (b).

coordinates—or conversely—by means of the Lorentz transformation equations (Eq. A-1), thus verifying the graphical relationships displayed in the Minkowski diagram.

In using the Minkowski diagram it is almost as if the rectangular grid of coordinate lines of $S$ (Fig. A-5a) became squashed toward the 45° bisecting line when the coordinate lines of $S'$ are put on the same graph (Fig. A-5b). In more formal language, we say that the Lorentz transformation equations transform an orthogonal (perpendicular) reference frame into a nonorthogonal one. Note that as $\beta \rightarrow 1$, corresponding to $v \rightarrow c$, the angle $\phi$ in Fig. A-5b ($= \tan^{-1} \beta$) approaches 45°, thus compressing the $S'$-frame coordinate space into a thinner and thinner wedge of the $S$-frame coordinate space. Alternatively, as $\beta \rightarrow 0$, corresponding to an approach to

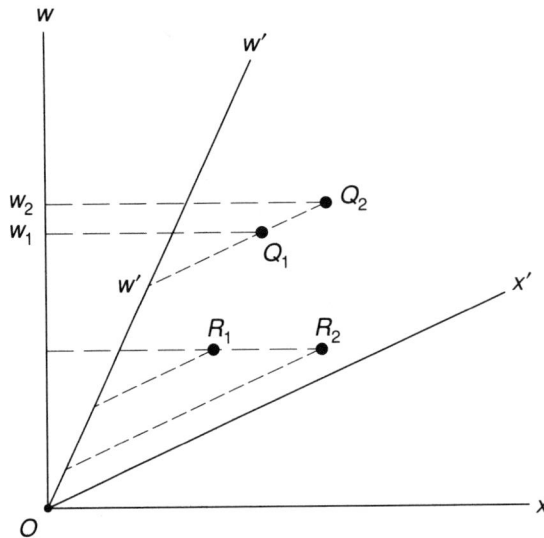

**FIGURE A-6.** Showing the relativity of simultaneity.

classical conditions, the angle $\phi$ between corresponding $S$ and $S'$ axes becomes very small. Even for a speed as high as that of a typical earth satellite ($\sim$17,000 mi/h), we note that $\beta = 2.5 \times 10^{-5}$, which yields a value of only 0.0015° for $\phi$; relativistic mechanics is not much different from classical mechanics in these circumstances.

## A-3 Simultaneity, Contraction, and Dilation

Now we can easily show the relativity of simultaneity. As measured in $S'$, two events will be simultaneous if they have the same time coordinate $w'$. Hence, if the events lie on a line parallel to the $x'$ axis, they are simultaneous to $S'$. In Fig. A-6, for example, events $Q_1$ and $Q_2$ are simultaneous in $S'$; they obviously are not simultaneous in $S$, occurring at different times $w_1$ and $w_2$ there. Similarly, two events $R_1$ and $R_2$, which are simultaneous in $S$, are separated in time in $S'$.

As for the space contraction, consider Fig. A-7a. Let a meter stick be at rest in the $S$ frame, its end points being at $x = 3$ and $x = 4$, for example. As time goes on, the world line of each end point traces out a vertical line parallel to the $w$ axis. The length of the stick is defined as the distance between the end points measured simultaneously. In $S$, the rest frame, the length is the distance in $S$ between the intersections of the world lines with the $x$ axis, or any line parallel to the $x$ axis, for these intersecting points represent simultaneous events in $S$. The rest length is one meter. To get the length of the stick in $S'$, where the stick moves, we must obtain the distance in $S'$ between end points measured simultaneously. This will be the separation in $S'$ of the intersections of the world lines with the $x'$ axis, or any line parallel to the $x'$ axis, for these intersecting points represent simultaneous events in $S'$. The length of the (moving) stick is clearly less than one meter in $S'$ (see Fig. A-7a).

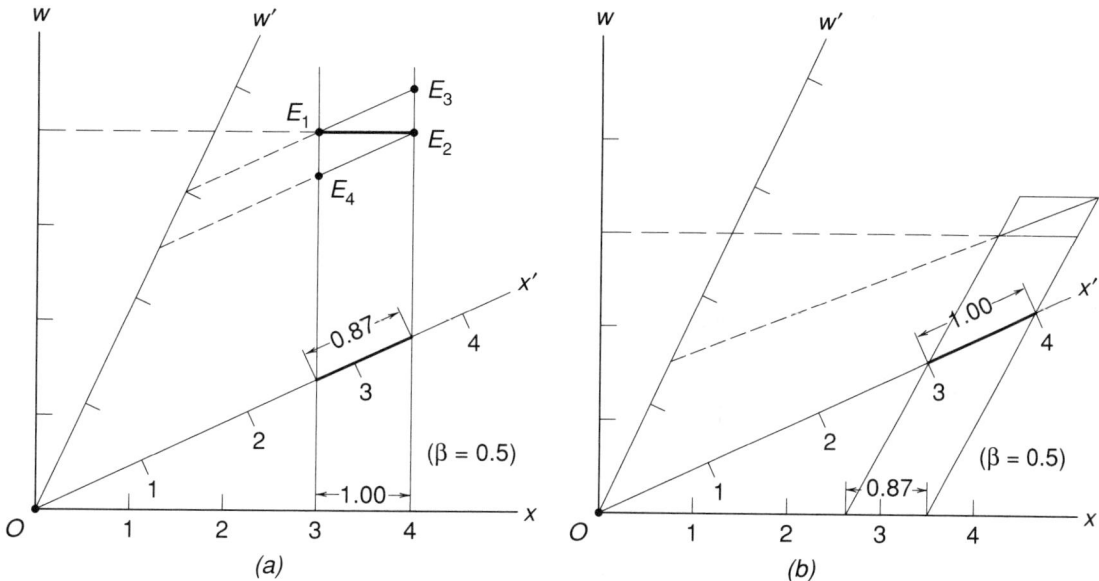

**FIGURE A-7.** Showing the space contraction, (a), and its reciprocal nature, (b).

Notice how very clearly Fig. A-7a reveals that it is a disagreement about the simultaneity of events that leads to different measured lengths. Indeed, the two observers do not measure the same pair of events in determining the length of a body (for example, the S observer uses $E_1$ and $E_2$, say, whereas the S' observer would use $E_1$ and $E_3$, or $E_2$ and $E_4$) for events that are simultaneous to one inertial observer are *not* simultaneous to the other. We should also note that the x' coordinate of each end point decreases as time goes on (simply project from successive world-line points parallel to w' onto the x' axis), consistent with the fact that the stick that is at rest in S moves towards the left in S' .

The reciprocal nature of this result is shown in Fig. A-7b. Here, we have a meter stick at rest in S', and the world lines of its end points are parallel to w' (the end points are always at x' = 3 and x' = 4, say). The rest length is one meter. In S, where the stick moves to the right, the measured length is the distance in S between intersections of these world lines with the x axis, or any line parallel to the x axis. The length of the (moving) stick is clearly less than one meter in S (Fig. A-7b).

It remains now to demonstrate the time-dilation result geometrically. For this purpose consider Fig. A-8. Let a clock be at rest in frame S, ticking off units of time there. The solid vertical line in Fig. A-8, at x = 2.3, is the world line corresponding to such a single clock. $T_1$ and $T_2$ are the events of ticking at w (= ct) = 2 and w (= ct) = 3, the time interval in S between ticks being unity. In S', this clock is moving to the left so that it is at a different place there each time it ticks. To measure the time interval between events $T_1$ and $T_2$ in S', we use two different clocks, one at the location of event $T_1$ and the other at the location of event $T_2$. The difference in reading of these clocks in S' is the difference in times between $T_1$ and $T_2$ as measured in S'. From the graph, we see that this interval is greater than unity.

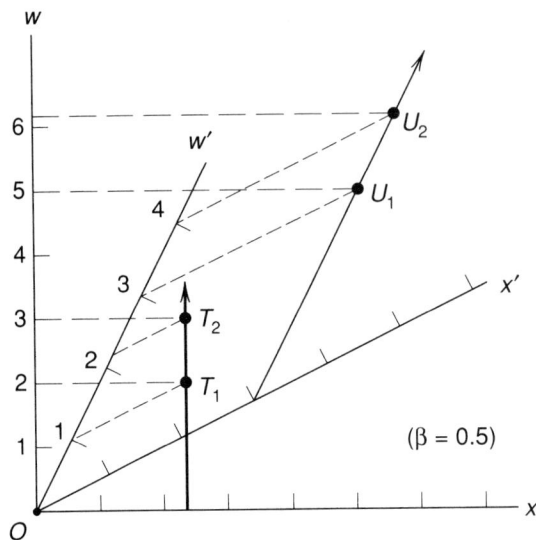

**FIGURE A-8.** Showing time dilation.

Hence, from the point of view of $S'$, the moving $S$ clock appears slowed down. During the interval that the $S$ clock registered unit time, the $S'$ clock registered a time greater than one unit.

The reciprocal nature of the time-dilation result is also shown in Fig. A-8. You should construct the detailed argument. Here a clock at rest in $S'$ emits ticks $U_1$ and $U_2$ separated by unit proper time. As measured in $S$, the corresponding time interval exceeds one unit.

## A-4 The Time Order and Space Separation of Events

We can also use the geometric representation of spacetime to gain further insight into the concepts of simultaneity and the time order of events that we discussed in Chapter 2. Consider the shaded area in Fig. A-9, for example. Through any point $P$ in this shaded area, bounded by the world lines of light waves, we can draw a $w'$ axis from the origin; that is, we can find an inertial frame $S'$ in which the events $O$ and $P$ occur at the same place ($x' = 0$) and are separated only in time.* As shown in Fig. A-9, event $P$ follows event $O$ in time (it comes later on $S'$ clocks), as is true wherever event $P$ is in the upper half of the shaded area. Hence, events in the upper half (region 1 on Fig. A-10) are absolutely in the future relative to $O$, and this region is called the Absolute Future. If event $P$ is at a spacetime point in the lower half of the shaded area (region 2 on Fig. A-10), then $P$ will precede event $O$ in time. Events in the lower half are absolutely in the past relative to $O$, and this region is called the Absolute Past. In the shaded regions, therefore, there is a definite time order of

---

*We cannot draw an $x'$ axis through points such as $P$ in Fig. A-9 because the angle $\phi$ in Fig. A-2 would then exceed 45°, which requires that $\beta > 1$ (or, equivalently, that $v > c$). For the same reason, we cannot draw a $w'$ axis through points such as $Q$ in Fig. A-9.

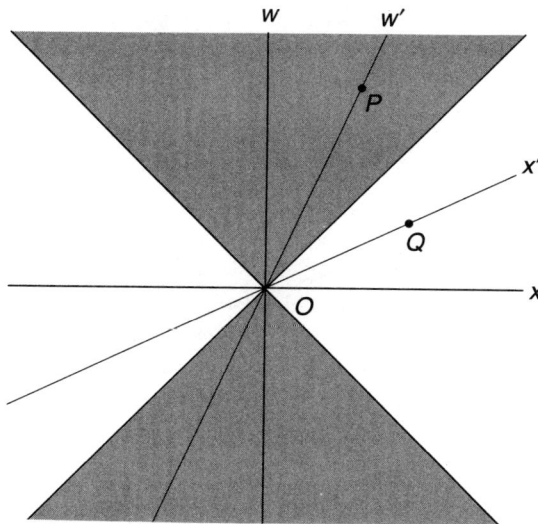

**FIGURE A-9.**  The time order and space separation of events.

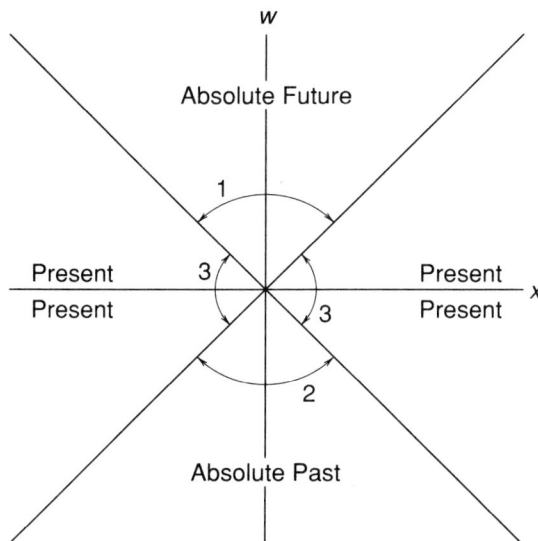

**FIGURE A-10.**  Location in time of events relative to the origin.

events relative to $O$, for we can always find a frame in which $O$ and $P$ occur at the same place; a single clock will determine absolutely the time order of the event at this place.

Consider now the unshaded regions of Fig. A-9. Through any point $Q$ we can draw an $x'$ axis from the origin; that is, we can find an inertial frame $S'$ in which the events $O$ and $Q$ occur at the same time ($w' = ct' = 0$) and are separated only in space. We can always find an inertial frame in which events $O$ and $Q$ appear to be simultaneous for spacetime points $Q$ that are in the unshaded regions (region 3 of Fig. A-10), so that this region is called the Present. In other inertial frames, of course, $O$ and $Q$ are not simultaneous, and there is no absolute time order of these events but a relative time order, instead.

If we ask about the space separation of events, rather than their time order, we see that events in the Present are absolutely separated from $O$, whereas those in the Absolute Future or Absolute Past have no definite space order relative to $O$. Indeed, region 3 (Present) is said to be "spacelike" whereas regions 1 and 2 (Absolute Past or Future) are said to be "timelike." That is, a world interval such as $OQ$ is spacelike and a world interval such as $OP$ is timelike.

The geometric considerations that we have presented are connected with the invariant nature of the spacetime *interval*, described in Section 2.3. As presented there, the interval involves a pair of events. For our purposes we can choose as one universal member of this pair the standard reference event represented by point $O$ in Fig. A-9. It corresponds to the coincidence in time of the origins of the two reference frames, $S$ and $S'$, and has the spacetime coordinates $x = w = 0$ and $x' = w' = 0$. The other member of the event pair can then be a generalized event represented by points such as $P$ or $Q$ in Fig. A-9. In this way we can associate the spacetime interval with $P$ and $Q$ alone, and can write (from Eq. 2-16, recalling that $w = ct$),

$$s^2 = w^2 - x^2 = w'^2 - x'^2. \tag{A-2}$$

We have seen that $s^2$, which has the same numerical value in all reference frames, can be either positive, negative, or zero, depending on the relative magnitudes of $w$ and $x$ (or of $w'$ and $x'$). If $w > x$, as it is for points such as $P$ in Fig. A-9, then $s^2$ is positive and $s$ is a real quantity; we write it as $c\tau$, where $\tau$ is the *proper time interval* associated with the event pairs such as $OP$; see Eq. 2-17. If $w < x$, as it is for points such as $Q$ in Fig. A-9, then $-s^2$ is a positive quantity; we call its square root $\sigma$, the *proper distance interval* for the event pairs such as $OQ$. We have then two relations,

$$c^2\tau^2 = w^2 - x^2 \tag{A-3a}$$

and

$$\sigma^2 = x^2 - w^2. \tag{A-3b}$$

Now consider Fig. A-10. In regions 1 and 2 we have spacetime points for which $w > x$, so the proper time is a real quantity, $c^2\tau^2$ being positive; see Eq. A-3a. In regions 3 we have spacetime points for which $x > w$, so the proper distance $\sigma$ is a real quantity; see Eq. A-3b. Hence either $\tau$ or $\sigma$ is real for any two events (that is, the event at the origin and the event elsewhere in spacetime) and either $\tau$ or $\sigma$ may be called the spacetime interval between the two events. When $\tau$ is real the interval is called "timelike"; when $\sigma$ is real the interval is called "spacelike." Because $\sigma$ and $\tau$ are invariant properties of two events, it does not depend at all on what inertial frame is used to specify the events whether the interval between them is spacelike or timelike.

In the spacelike region we can always find a frame $S'$ in which the two events are simultaneous, so that $\sigma$ can be thought of as the spatial interval between the events in that frame. (That is, $\sigma^2 = x^2 - w^2 = x'^2 - w'^2$. But $w' = 0$ in $S'$, so $\sigma = x'$.) In the timelike region we can always find a frame $S'$ in which the two events occur at the same place, so that $\tau$ can be thought of as the time interval between the events in that frame. [That is, $\tau^2 = t^2 - (x^2/c^2) = t'^2 - (x'^2/c^2)$. But $x' = 0$ in $S'$, so $\tau = t'$.]

What can we say about points on the 45° lines? For such points, $x = w$. Therefore, the proper time interval between two events on these lines vanishes, for $c^2\tau^2 = w^2 - x^2 = 0$ if $x = w$. We have seen that such lines represent the world lines of light rays and give the limiting velocity ($v = c$) of relativity. On one side of these 45° lines (shaded regions in Fig. A-9), the proper time interval is real; on the other side (unshaded regions), it is imaginary. An imaginary value of $\tau$ would correspond to a velocity in excess of $c$. But no signals can travel faster than $c$. All this is relevant to an interesting question that can be posed about the unshaded regions.

In this region, which we have called the Present, there is no absolute time order of events; event $O$ may precede event $Q$ in one frame but follow event $Q$ in another frame. What does this do to our deep-seated notions of cause and effect? Does relativity theory negate the causality principle? To test cause and effect, we would have to examine the events at the same place so that we could say absolutely that $Q$ followed $O$, or that $O$ followed $Q$, in each instance. But in the Present, or spacelike, region these two events occur in such rapid succession that the time difference is less than the time needed by a light ray to traverse the spatial distance between two events. We cannot fix the time order of such events absolutely, for no signal can travel from one event to the other faster than $c$. In other words, no frame of reference exists with respect to which the two events occur at the same place; thus, we simply cannot test causality for such events even in principle. Therefore, there is no violation of the law of causality implied by the relative time order of $O$ and events in the spacelike region. We can arrive at this same result by an argument other than this operational one. If the two events, $O$ and $Q$, are related causally, then they must be capable of interacting physically. But no physical signal can travel faster than $c$, so events $O$ and $Q$ cannot interact physically. Hence, their time order is immaterial, for they cannot be related causally. Events that can interact physically with $O$ are in regions other than the Present. For such events, $O$ and $P$, relativity gives an unambiguous time order. Therefore, relativity is completely consistent with the causality principle.

## QUESTIONS AND PROBLEMS

1. **Interpreting events on a spacetime diagram (I).** Draw a spacetime diagram and on it locate an event $P$ whose coordinates are $x = 450$ m and $t = 1.00$ μs ($w = ct = 300$ m). With respect to the standard reference event $O$ at the origin, (a) does $P$ represent an event in the future? The present? The past? (b) Is the interval $OP$ spacelike? Timelike?

Lightlike? (c) What proper time interval is associated with OP? (d) What proper space interval? (e) Can you find another frame S' for which the events OP would occur at the same time? If so, draw the spacetime axes of that frame on your diagram and give the speed parameter β of the frame. (f) Can you find another frame S'' in which the events OP would occur at the same place? If so, draw the spacetime axes of that frame on your diagram and give its speed parameter β.

2. **Interpreting events on a spacetime diagram (II).** Solve Problem 1 for an event whose coordinates are x = 450 m and t = 1.50 μs, and for an event whose coordinates are x = 450 m and t = 1.00 μs. Plot both events on the spacetime diagram of Fig. A-10 and compare.

3. **Present or Absolute Future?** Consider two events, both of which have x = 1 in, say, Fig. A-4. Event A has w = 0.9 and event B has w = 1.1. Comparison with Fig. A-10 shows that, with respect to the reference event O at the origin, the first of these events would be classified as Absolute Future and the second as Present. However, they both seem to occur in the future from the point of view of observer S. If Present means "right now," then neither of these events seems to qualify. It also seems clear that these two events could differ as slightly as you please; one, for example, could have w = 0.99999 and the other w = 1.00001, and a fundamental difference would still remain between them. Can you identify and clarify this difference?

★4. **Calibrating the axes.** (a) Let d represent the length (measured with a ruler) of a unit interval on the x (or the w) axis of Fig. A-4 and let d' represent the corresponding quantity for the x' (or the w') axis of that figure. Show that

$$\frac{d'}{d} = \frac{\gamma}{\cos \phi} = \sqrt{\frac{1 + \beta^2}{1 - \beta^2}}.$$

(b) Evaluate this ratio for β = 0.50 (for which Fig. A-4 is drawn) and verify (using a ruler) that the axes in that figure are calibrated correctly.

5. **Changing the scales of a spacetime diagram.** (a) Redraw the spacetime diagram of Fig. A-4 but, in place of the dimensionless scale factor of unity, take 200 m as the unit scale distance. On this diagram, locate an event P whose spacetime coordinates, as determined by observer S, are x = 800 m and t = 1.00 μs. Determine the spacetime coordinates of this event as determined by observer S' (b) directly from your diagram and (c) using the Lorentz transformation equations.

6. **Learning about the time axes.** In the spacetime (or Minkowski) diagram of Fig. A-2, the w axis, from the point of view of observer S, represents the world line of a particle resting at the origin of the S frame. Identify on the diagram the world line (a) of the S origin from the point of view of S'; (b) of the S' origin from the point of view of S; (c) of the S' origin from the point of view of S'. (d) Write down the equations of all four of these world lines, in coordinates appropriate to the observer, using the Lorentz transformation equations (see Eqs. A-1) as needed.

7. **Learning about the position axes.** In the spacetime diagram of Fig. A-2, the x axis, from the point of view of observer S, is made up of points at each of which there is a clock, fixed in the S frame, that reads w = 0. (a) What times would observer S' read on these same S clocks? (b) Identify the locus of points each of which contains a clock fixed in the S' frame that reads w' = 0. (c) What times would the S observer read on those S' clocks?

8. **S and S' watch a clock.** In Fig. A-4, consider a clock at rest at the origin of the S' frame and consider an event corresponding to a reading of "3" (as seen by observer S') on

that clock. What reading will observer $S$ (who uses his own clocks) record for this event? Solve by direct measurement from the spacetime diagram and also by use of the Lorentz transformation equations. Recall that $\beta = 0.50$ for the conditions of Fig. A-4.

9. **$S$ and $S'$ measure a rod.** In Fig. A-4, consider a rod 2.00 units long at rest along the $x'$ axis of the $S'$ reference frame, one end of the rod being at the origin of that frame. Both observers, $S$ and $S'$, measure the length of the rod. What values do they find? Solve by direct measurement from the spacetime diagram and also by use of the Lorentz transformation equations. Show on the spacetime diagram the two events that are used by each observer in the measuring process.

10. **You can't get there that fast.** Let the departure of a plane from Boston be an event whose coordinates are $x = 0$ and $w = ct = 0$. Let a second event be the arrival of that plane in Seattle. Plot these two events qualitatively on the spacetime diagram of Fig. A-10. (a) Can you find a second frame $S'$ in which these two events are simultaneous? If so, describe that frame. (b) Can you find a frame in which these two events occur at the same place? If so, describe *that* frame. (c) Is the interval associated with these two events spacelike or timelike?

11. **Three reference frames.** A system $S'$ moves to the right relative to $S$ at a speed of $0.60c$ and another system $S''$ moves to the right relative to $S$ at a speed $0.35c$. (a) Using the Minkowski diagram, find the velocity of $S''$ relative to frame $S'$. (b) Repeat, with $S''$ moving to the right at a speed of $0.50c$. (*Hint:* Construct lines of constant $x'$ and $t'$ on the diagram. Using this gridwork of lines, find the slope of the world line for $\beta = 0.35$ and for $\beta = 0.50$).

12. **An event viewed from three reference frames.** (a) Redraw the spacetime diagram of Fig. A-4 and superimpose on it a set of axes corresponding to frame $S''$, which is moving in the positive $x$ direction with a speed $0.80c$. Calibrate the $x''$-$w''$ axes described in Problem 4. (b) Consider an event, represented by $P$ in Fig. A-4, whose coordinates in $S$ are $x = 3.0$ and $w = 2.5$. Find the coordinates of this event in frame $S'$ and in frame $S''$, both directly from the diagram and by use of the Lorentz transformation equations.

13. **A collision on a spacetime diagram.** A particle of mass $m$ is at rest at $x = 3$ m along the $x$ axis of a coordinate system. A second particle, of mass $2m$, passes through the origin at $t = 0$ and moves toward the first particle with a speed of 1 m/s. The two particles then undergo a head-on collision, both particles moving forward along the $x$ axis after the collision. (a) Show that, according to classical physics, after the collision the initially moving particle moves forward with a speed of $\frac{1}{3}$m/s and the initially resting particle does so with a speed of $\frac{4}{3}$m/s. (b) Draw the world lines for these particles on a spacetime diagram, for an interval encompassing the collision. (c) Draw on the same diagram the world line representing the motion of the center of mass of the colliding particles.

14. **Using spacetime diagrams (I).** Read again Problems 31 and 32 of Chapter 2. (a) Draw a world diagram for the problem, including on it the world lines for observers $A$, $B$, $C$, $D$, and $E$. Label the points $AD$, $BD$, $AC$, $BC$, and $EC$. (b) Show, by means of the diagram, that the clock at $A$ records a shorter time interval for the events $AD$, $AC$ than do clocks at $D$ and $C$. (c) Show that, if observers on the cart try to measure the length $DC$ by making simultaneous markings on a measuring stick in their frame, they will measure a length shorter than the rest length $DC$. Explain this result in terms of simultaneity, using the diagram. For convenience, take $v = \frac{1}{2}c$.

15. **Using spacetime diagrams (II).** Do Example 4, Chapter 2, by means of a Minkowski

diagram. Check your results by calculating the invariants $c^2\tau^2$ or $\sigma^2$.

16. **Using spacetime diagrams (III).** Do Example 5, Chapter 2, by means of a Minkowski diagram. Check your results by calculating the invariants $c^2\tau^2$ or $\sigma^2$.

17. **Calibration by hyperbolas.** Equations A-3 represent the equations of hyperbolas, each of which has two branches. If we choose both $\sigma$ and $c\tau$ equal to one scale unit on the $x$ and $w$ axes of a spacetime diagram, we can write, for the equations of these hyperbolas,

$$w^2 - x^2 = 1 \quad \text{(timelike region)}$$

and

$$x^2 - w^2 = 1 \quad \text{(spacelike region).}$$

Figure A-11 shows these hyperbolas; they represent but one typical family of an infinite set of hyperbolas, corresponding to different choices for $\sigma$ and for $c\tau$ in Eqs. A-3. Figure A-12 shows the upper right quadrant of Fig. A-11, with the spacetime axes of two refer-

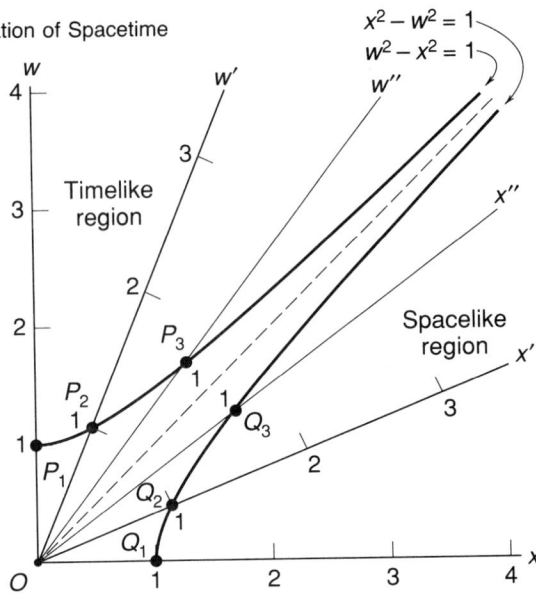

**FIGURE A-12.** Problem 17; the first quadrant in detail.

ence frames $S'$ ($\beta = 0.50$) and $S''$ ($\beta = 0.80$) drawn in.

(a) Prove that the hyperbola branches in Fig. A-11 approach the 45° lightlike lines asymptotically. (b) Sketch in roughly the curves that would correspond to a choice of $\sigma = 2$ in Eqs. A-3 and to a choice of $c\tau = 2$. (c) With respect to the standard reference event $O$, what proper time interval (in terms of $w$) can you assign to events such as $J$, $K$, $L$, and $M$ in Fig. A-11? What proper space interval? (d) In Fig. A-11, convince yourself that the point representing the intersection of the $w$ axis and the upper branch of the timelike hyperbola corresponds to a clock at rest at the origin of the $S$ frame and reading "1." Other points on this hyperbola branch read times greater than "1"; where are these clocks located? Are they fixed in the $S$ frame, or are they moving? What does it mean when we say that, though one of these clocks may read, say, "1.5," its *proper* time is still "1"? (e) In Fig. A-12 we have seen that $P_1$ represents a clock at the origin of the

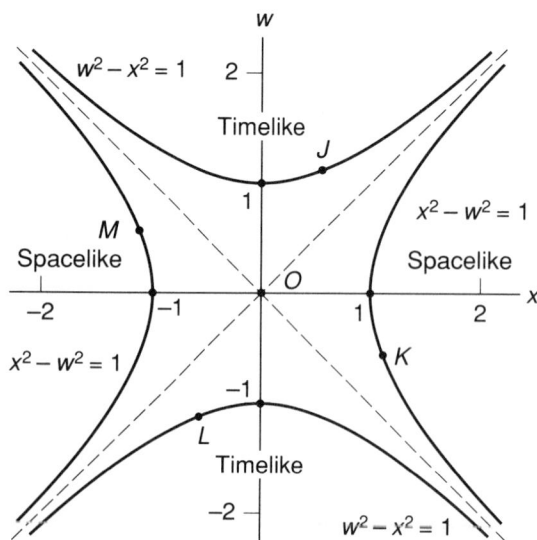

**FIGURE A-11.** Problem 17; calibration by hyperbolas.

*S* frame reading "1." Can you see that $P_2$ represents a clock resting at the origin of the *S'* frame and also reading "1"? What does $P_3$ represent? Can you see how the hyperbolas of Figs. A-11 and A-12 can be used to establish scales for the axes of a given reference frame? (The hyperbolas are often called "calibration curves," for this reason.) (*f*) Analyze the spacelike hyperbola branch shown in Fig. A-12 in physical terms, following the pattern outlined above for the timelike branch. In particular, to what do events $Q_1$, $Q_2$, and $Q_3$ correspond physically?

## REFERENCES

**1.** A. EINSTEIN, H. MINKOWSKI, H. A. LORENTZ, and H. WEYL, *The Principle of Relativity: A Collection of Original Memoirs on the Special and General Theory of Relativity,* notes by A. Sommerfeld (Dover, New York, 1952.).

**2.** HERMANN MINKOWSKI, "Space and Time" (a translation of an address given September 21, 1908), in *The Principle of Relativity* (Dover, New York).

# The Twin Paradox

*If we placed a living organism in a box . . . one could arrange that the organism, after any arbitrary lengthy flight, could be returned to its original spot in a scarcely altered condition, while corresponding organisms which had remained in their original positions had already long since given way to new generations. For the moving organism the lengthy time of the journey was a mere instant, provided the motion took place with approximately the speed of light.*

*Albert Einstein (1911)*

In the above statement Einstein describes what has come to be called the twin paradox or the clock paradox [1]. If the stationary organism is a man and the traveling one is his twin, then the traveler returns home to find his twin brother much aged compared to himself. The paradox centers around the contention that, in relativity, either twin could regard the other as the traveler, in which case each should find the other younger—a logical contradiction. This contention assumes that the twins' situations are symmetrical and interchangeable, an assumption that is not correct. Furthermore, the accessible experiments have been done and support Einstein's prediction. In succeeding sections, we look with some care into the many aspects of this problem.

## B-1 The Elapsed Proper Time Depends on the Route

Figure B-1*a* shows the world lines of three particles, their motions being confined to the *x* axis of an inertial reference frame. Particle 1 is at rest on the *x* axis, its world line being a simple vertical line. Particle 2 is moving along this axis in the direction of increasing *x*, its (constant) speed *v* being $dx/dt = c\, dx/(c\, dt) = c \tan \theta$, where $\theta$ is the angle made by the world line of this particle with the vertical. A reference frame moving with particle 2 would thus be an inertial frame. Particle 3 is in accelerated motion along the *x* axis, its speed *v* being, in general, different for every position of the particle. A frame moving with *this* particle would *not* be an inertial frame.

Let us assume that particle 3 in Fig. B-1*a* is, in fact, a clock, and let us consider the problem of calculating the elapsed proper time on this clock as it travels from one point to another. We can assume that this traveling clock is in an inertial frame

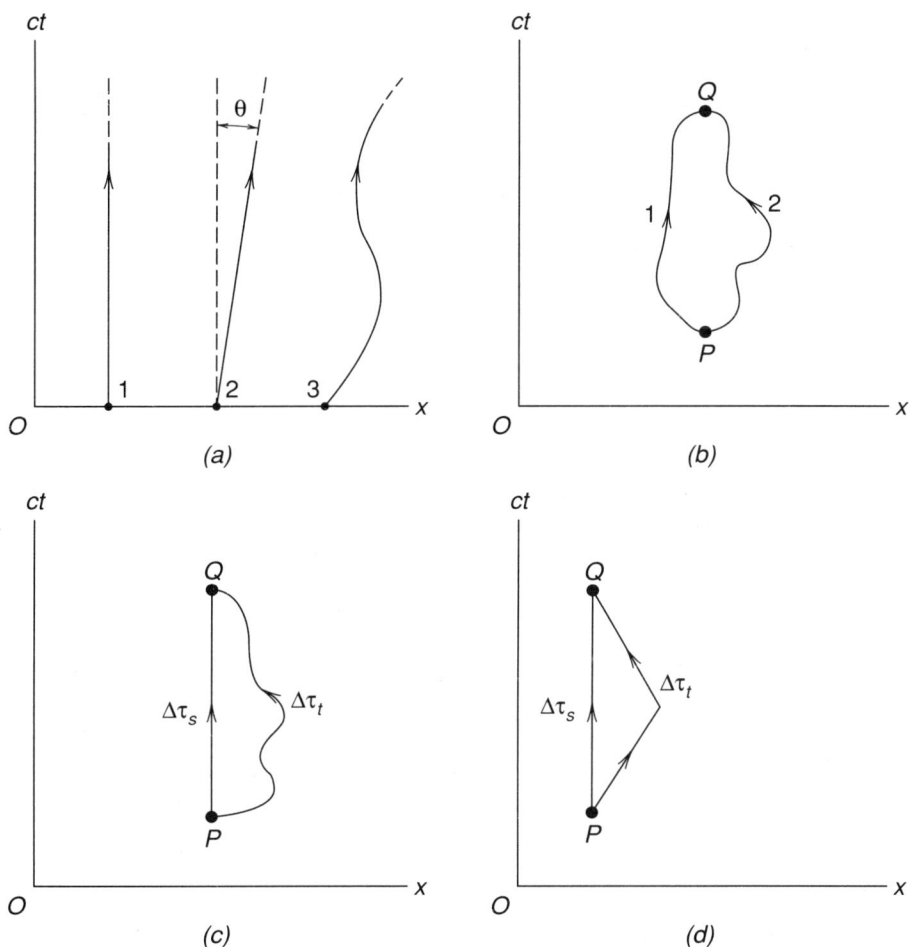

**FIGURE B-1.** World lines for various motions in an inertial frame.

only for differential elements of its path. We can compute the elapsed proper time $d\tau$ for such an element from Eq. 2-17, which we write in differential form as

$$(c\ d\tau)^2 = (c\ dt)^2 - (dx)^2$$

or

$$d\tau = \sqrt{(dt)^2 - (\frac{dx}{c})^2}. \tag{B-1}$$

The total elapsed proper time between any two points would then be the integral of this quantity between those points.

Consider now Fig. B-1*b*, in which the world lines of two particles, each with a clock attached, start from point $P$ and reconvene at $Q$. For either path the proper time interval—that is, the elapsed time on the particle's clock—between $P$ and $Q$ is given by (see Eq. B-1)

$$\Delta\tau = \int_P^Q d\tau = \int_P^Q \sqrt{(dt)^2 - (\frac{dx}{c})^2}. \qquad (B\text{-}2)$$

Both $dt$ and $dx$ in the above are differential spacetime path elements as measured by the observer in the inertial reference frame of Fig. B-1$b$. We are not surprised that the two paths shown in this figure differ as far as $x$ is concerned (odometer readings), and we have learned not to be surprised that clock readings vary in much the same way. Simple inspection of Eq. B-2 shows that the quantity depends not only on the initial and final points but also on the path taken between them.

In Fig. B-1$c$ we let one of these paths be a straight line, corresponding to the simple passage of time for a stationary particle; the other path remains arbitrary. From Eq. B-2 we have, for the straight path,

$$\Delta\tau_s = \int_P^Q \sqrt{(dt)^2 - \left(\frac{dx}{c}\right)^2} = \int_P^Q dt = t_Q - t_P,$$

in which the subscript on $\Delta\tau$ refers to the stationary clock. In such a case $dx$ is zero along the path, and the proper time coincides with the time interval, $t_Q - t_P$, recorded by the stationary clocks of the inertial reference frame. Along the second world line, however, the elapsed power time is

$$\Delta\tau_t = \int_P^Q \sqrt{(dt)^2 - \left(\frac{dx}{c}\right)^2},$$

in which the subscript refers to the traveling clock; we see that $\Delta\tau_t$ will *not* equal $\Delta\tau_s$. In fact, since $(dx)^2$ is always positive, we find that

$$\Delta\tau_t < \Delta\tau_s. \qquad (B\text{-}3)$$

The clocks will read different times when brought back together, the traveling clock running behind (recording a smaller time difference than) the stay-at-home clock. Figure B-1$d$ is a special case of Fig. B-1$c$ in that the traveling clock moves with constant velocity over most of its path, its motion being accelerated only near its "turnaround point." Note that, although the turnaround may occupy only a small fraction of the total travel time, it is vitally necessary to the motion if the two clocks are to reconvene.

We have noted that the reference frame whose axes are drawn in Fig. B-1 is an inertial frame. The motion of the traveling clock is represented in this frame by a curved world line, for this clock undergoes accelerated motion rather than motion with uniform velocity. It could not return to the stationary clock, for example, without reversing its velocity. The special theory of relativity can predict the behavior of accelerated objects as long as, in the formulation of the physical laws, we take the view of the inertial (unaccelerated) observer. This is what we have done so far. A frame attached to the clock traveling along its round-trip path would not be an inertial frame. We could reformulate the laws of physics so that they have the same form for accelerated (noninertial) observers—this is the program of general relativity theory—but it is unnecessary to do so to explain the twin paradox. All we wish to point out here is that the situation is *not* symmetrical with respect to the clocks (or twins); one is always in a single inertial frame and the other is not.

# **B-2** Spacetime Diagram of the Twin Paradox

In our earlier discussions of time dilation, we spoke of "moving clocks running slow." What is meant by that phrase is that a clock moving at a constant velocity **u** relative to an inertial frame containing synchronized clocks will be found to run slow by the factor $\sqrt{1 - u^2/c^2}$ *when timed by those clocks.* That is, to time a clock moving at constant velocity relative to an inertial frame, we need at least *two* synchronized clocks in that frame. We found this result to be reciprocal in that a single $S'$ clock is timed as running slow by the many $S$ clocks, and a single $S$ clock is timed as running slow by the many $S'$ clocks.

The situation in the twin paradox is different. If the traveling twin traveled always at a constant speed in a straight line, he would never get back home. And each twin would indeed claim that the other's clock runs slow compared to the synchronized clocks in his own frame. To get back home—that is, to make a round trip—the traveling twin would have to change his velocity. What we wish to compare in the case of the twin paradox is a single moving clock with a *single* clock at rest. To do this we must bring the clocks into coincidence twice—they must come back together again. It is not the idea that we regard one clock as moving and the other at rest that leads to the different clock readings, for if each of two observers seems to the other to be moving at constant speed in a straight line, they cannot absolutely assert who is moving and who is not. Instead, it is because one clock has *changed* its velocity and the other has not that makes the situation unsymmetrical.

Now you may ask how the twins can tell who has changed his velocity. This is clearcut. Each twin can carry an accelerometer. If he changes his speed or the direction of his motion, the acceleration will be detected. We may not be aware of an airplane's motion, or a train's motion, if it is one of uniform velocity; but let it move in a curve, rise and fall, speed up or slow down, and we are our own accelerometer as we get thrown around. Our twin on the ground watching us does not experience these feelings—his accelerometer registers nothing. Hence, we can tell the twins apart by the fact that the one who makes the round-trip experiences and records accelerations whereas the stay-at-home does not.

A numerical example, suggested by C. G. Darwin [2], is helpful in fixing the ideas. We imagine that, on New Year's Day, Bob leaves his twin brother Dave, who is at rest on a spaceship, fires rockets that get him moving at a speed of $0.8c$ relative to Dave, and by his own clock travels away at this constant speed toward a distant star, which he reaches after three years of travel. He then fires more powerful rockets that exactly reverse his motion and gets back to Dave after another three years by his clock. By firing rockets a third time, he comes to rest beside Dave and compares clock readings. Bob's clock says he has been away for six years (the $\Delta\tau_t$ of Eq. B-3), but Dave's clock says that ten years have elapsed (the $\Delta\tau_s$ of Eq. B-3). Let us see how this comes about.

First, we can simplify matters by ignoring the effect of the accelerations on the traveling clock. Bob can turn off his clock during the three acceleration periods, for example. The error thereby introduced can be made very small compared to the total time of the trip, for we can make the trip as far and as long as we wish without

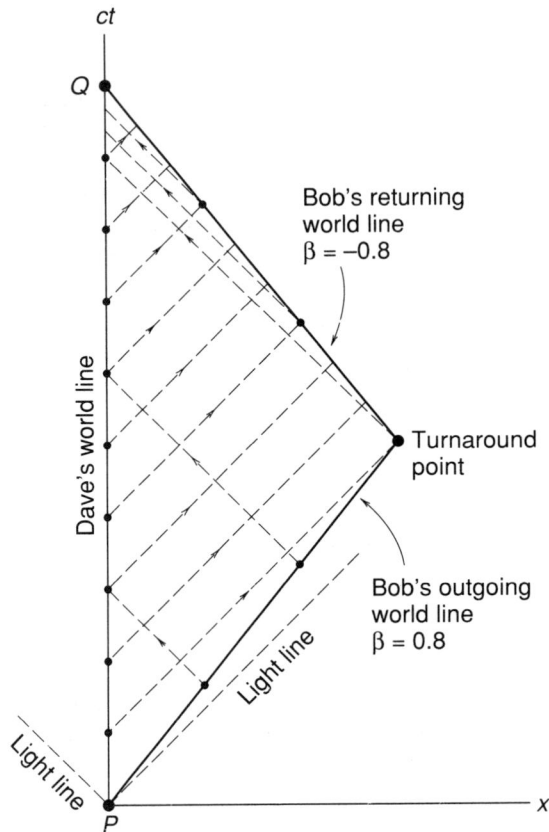

**FIGURE B-2.**    Spacetime diagram of the twin paradox.

changing the acceleration intervals. It is the total time that is at issue here in any case.* We do not destroy the asymmetry, for even in the ideal simplification of Fig. B-2 (where the world lines are straight lines rather than curved ones), Dave is always in one inertial frame whereas Bob is definitely in two different inertial frames—one going out (0.8c) and another coming in (−0.8c).

Let the spaceships be equipped with identical clocks that send out light signals at one-year intervals. Dave receives the signals arriving from Bob's clock and records them against the annual signals of his own clock; likewise, Bob receives the signals from Dave's clock and records them against the annual signals of his clock.

In Fig. B-2, Dave's world line is straight along the ct-axis; he is at $x = 0$ and we mark off ten years (in terms of ct), a dot corresponding to the annual New Year's

---

*An analogy is that the total distance traveled by two drivers between the same two points, one along the hypotenuse of a right triangle and the other along the other two sides of the triangle, can be quite different. One driver always moves along a straight line, whereas the other makes a right turn to travel along two straight lines. We can make the distance between the two points as long as we wish without altering the fact that only one turn must be made. The difference in mileage traveled by the drivers certainly is not acquired at the turn that one of them makes.

Day signal of his clock. Bob's world line at first is a straight line inclined to the $ct$ axis, corresponding to a $ct'$ axis of a frame moving at $+0.8c$ relative to Dave's frame. We mark off three years (in terms of $ct'$), a dot corresponding to the annual New Year's Day signal of his clock. After three of Bob's years, he switches to another inertial frame whose world line is a straight line inclined to the $ct$ axis, corresponding to the $ct''$ axis of a frame moving at $-0.8c$ relative to Dave's frame. We mark off three years (in terms of $ct''$), a dot corresponding to the annual New Year's Day signal of his clock. Note the dilation of the time interval of Bob's clock compared to Dave's.

Now let us draw the light signals from Bob's clock on the spacetime diagram of Fig. B-2. Recall (see Fig. A-1) that such signals are drawn at 45° to the spacetime axes, corresponding to their speed of $c$. Thus from each dot on Bob's world line we draw such a 45°-line headed back to Dave on the line at $x = 0$. There are six signals, the last one emitted when Bob returns home to Dave. Likewise, the signals from Dave's clock are straight lines, from each dot on Dave's world line, inclined 45° to the axes and headed out to Bob's spaceship. We see that there are ten signals, the last one emitted when Bob returns home to Dave.

How can we confirm this spacetime diagram numerically? Simply by the Doppler effect. As the clocks recede from each other, the frequency $v$ of their signals is reduced from the proper frequency $v_0$ by the Doppler effect. From Eq. 2-30$b$ we thus have

$$\frac{v}{v_0} = \sqrt{\frac{1 - \beta}{1 + \beta}} = \sqrt{\frac{1 - 0.8}{1 + 0.8}} = \frac{1}{3}.$$

Hence, Bob receives the first signal from Dave after three of his years, just as he is turning back. Similarly, Dave receives messages from Bob on the way out once every three of his years, receiving three signals in nine years. As the clocks approach one another, the frequency $v$ of their signals is increased from the proper frequency $v_0$ by the Doppler effect. In this case (see Eq. 2-30$a$) we have

$$\frac{v}{v_0} = \sqrt{\frac{1 + \beta}{1 - \beta}} = \sqrt{\frac{1 + 0.8}{1 - 0.8}} = 3.$$

Thus, Bob receives nine signals from Dave in his three-year return journey. Altogether, Bob receives ten signals from Dave. Similarly, Dave receives three signals from Bob in the last year before Bob is home. Altogether, Dave receives six signals from Bob.

Figure B-3 shows the signal logs for Dave's and Bob's spaceships. Signals sent are indicated below the time axis in each case and signals received are shown above that axis. There is no disagreement about the signals: Bob sends six and Dave receives six; Dave sends ten and Bob receives ten. Everything works out, each seeing the correct Doppler shift of the other's clock and each agreeing to the number of signals that the other sent. The different total times recorded by the twins corresponds to the fact that Dave sees Bob recede for nine years and return in one year, although Bob both receded for three of his years and returned for three of his years. Dave's records will show that he received signals at a slow rate for nine years and at a rapid rate for one year. Bob's records will show that he received signals at

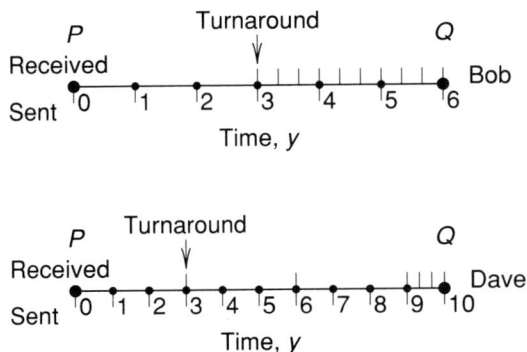

**FIGURE B-3.**  The signal logs for the twins.

a slow rate for three years and at a rapid rate for another three years. The essential asymmetry is thereby revealed by a Doppler effect analysis. When Bob and Dave compare records, they will agree that Dave's clock recorded ten years and Bob's recorded only six. Ten years have passed for Dave during Bob's six-year round trip.

## B-3 Some Other Considerations

Will Bob really be four years younger than his twin brother? Since for the word "clock" we could have substituted any periodic natural phenomena, such as heartbeat or pulse rate, the answer is yes. We might say that Bob lived at a slower rate than Dave during his trip, his bodily functions proceeding at the same slower rate as his physical clock. Biological clocks behave in this respect the same as physical clocks. There is no evidence that there is any difference in the physics of organic processes and the physics of the inorganic materials involved in these processes. If motion affects the rate of a physical clock, we expect it to effect a biological clock in the same way.

It is of interest to note the public acceptance of the idea that human life processes can be slowed down by refrigeration, so that a corresponding different aging of twins can be achieved by temperature differences. What is paradoxical about the relativistic case, in which the different aging is due to the difference in motion, is that since (uniform) motion is relative, the situation appears (incorrectly) to be symmetrical. But, just as the temperature differences are real, measurable, and agreed upon by the twins in the foregoing example, so are the differences in motion real, measurable, and agreed upon in the relativistic case—the changing of inertial frames, that is, the accelerations, are not symmetrical. The results are absolutely agreed upon.

Although there is no need to invoke general relativity theory in explaining the twin paradox, the student may wonder what the outcome of the analysis would be if we knew how to deal with accelerated reference frames. We could then use Bob's spaceship as our reference frame, so that Bob is the stay-at-home, and it would be Dave who, in this frame, makes the round-trip space journey. We would find that we must have a gravitational field in this frame to account for the accelerations that Bob feels and the fact that Dave feels no accelerations even though he makes a round trip. If, as required in general relativity, we then compute the frequency shifts of light in this gravitational field, we come to the same conclusion as in special relativity [3].

# B-4 Experimental Tests

Testing the conclusions we have reached with actual twins and clocks in spaceships moving with speeds close to the speed of light is, of course, more than can be managed at present. However, totally equivalent high-speed tests can be carried out using as clocks unstable elementary particles such as muons or pions or one of the hundreds of varieties of radioactive atoms available to us. At lower speeds—those of jet planes, for example—tests can be carried out with macroscopic atomic clocks, thanks to impressive improvements in the stability and time-keeping ability of such clocks.

In Section 2-7 we described the precise measurements of Kundig [Ref. 9, Chapter 2] on the transverse Doppler effect. This effect, as we noted in that section, is a direct measure of the time dilation, and we can use it to illustrate the twin paradox. From the point of view of the observer on the rotor axis (the stay-at-home twin), the absorbing foil on the perimeter of the rotor (the traveling twin) has a characteristic resonant absorption frequency that matches the source frequency only when the rotor is *not* turning. When the rotor *is* turning, the resonant frequency of the moving foil drops, just as predicted by Eqs. 2-32. Put another way, the round-trip twin ages less than his stay-at-home brother and (to within 1 percent) by exactly the amount predicted by relativity theory.

In 1968 a careful measurement of time dilation was reported from CERN (the European Nuclear Research Center, located near Geneva) in which laboratory-generated 1.18-GeV muons, for which the corresponding speed is $0.9966c$, served as high-speed traveling clocks [4]. These muons were constrained to circulate in an orbit 5.0 m in diameter in the muon storage ring in that laboratory. Thus, like the traveling twin (and also like the resonant absorbing foil in the experiment described above), they traverse a closed path and undergo (centripetal) acceleration during their journey. Their mean life for decay in flight can then be compared with the mean life observed when muons are brought to rest in an absorbing block. Many experiments give the accepted value of $2.200 \pm 0.0015$ µs for the decay of resting muons (the stay-at-home twin); the CERN experimenters measured $26.15 \pm 0.03$ µs for the mean decay time for the traveling muons (the traveling twin). This agrees within about 2 percent with the lifetime predicted for these traveling muons because of the time dilation, namely, $26.72$ µs. The time dilation phenomenon is universally accepted and is, in fact, turned to specific advantage in the design of certain high-energy particle experiments. As one high-energy physicist has written [5]: "We frequently transport beams of unstable particles over long distances such that no particles would be left without the help of Einstein's factor."

Refinements of atomic clocks have so improved the accuracy of timekeeping that time-dilation effects can be detected at speeds as low as those of jet planes. In

**TABLE B-1**
**Round-the-World Atomic Clocks** [6] (The numbers shown are time differences, in nanoseconds, with respect to reference clocks at the U.S. Naval Observatory.)

|  | Eastward | Westward |
|---|---|---|
| Predicted: | | |
|    Special relativity | $-184 \pm 18$ | $96 \pm 10$ |
|    General relativity | $144 \pm 14$ | $179 \pm 18$ |
| Net predicted: | $-40 \pm 23$ | $275 \pm 21$ |
| Observed: | $-59 \pm 10$ | $273 \pm 7$ |

October 1977, Joseph Hafele and Richard Keating [6] carried four cesium-beam atomic clocks around the world on commercial airline flights, " . . . to test Einstein's theory of relativity with macroscopic clocks." They took their clocks around once each way, that is, once eastward and once westward, comparing the traveling clocks to those that stayed at home at the Naval Observatory on the Earth, rotating (eastward) below them. The calculations must take into account not only the kinematic time-dilation effect (which is related only to the speed of the traveling clocks), but also relativistic frequency shifts associated with changes encountered in the strength of the earth's gravitational field (see Supplementary Topic C, Section C-2). Table B-1 shows the predictions of relativity theory along with the experimental findings. Hafele and Keating conclude: "There seems to be little basis for further arguments about whether clocks will indicate the same time after a round trip, for we find that they do not."

Today, when precision clocks move from one location to another, cumulative time corrections with respect to a stay-at-home clock are made routinely [7]. Such considerations enter, for example, when precision clocks are moved for comparison purposes between Washington, D.C., and the National Bureau of Standards Laboratory at Boulder, Colorado. Similarly, relativistic time-dilation effects must be considered in the design and operation of the Global Positioning System (GPS/NAVSTAR), a precision navigation system in which it is planned to employ 24 orbiting satellites.

## QUESTIONS AND PROBLEMS

1. **The shortest distance between two points is a straight line (?).** Comparison of Fig. B-1c and Eq. B-3 shows us that, in terms of elapsed proper time in units of $ct$ on a spacetime diagram, a straight line is not the shortest distance between two points but the *longest*. Is this statement still true if one of the two particles involved is not stationary (as in Fig. B-1c) but moves with constant speed? Draw a spacetime diagram to represent this situation.

2. **Einstein on the clock "paradox."** Einstein, in his first paper on the special theory of relativity, wrote the following: "If one of two synchronous clocks at A is moved in a closed curve with constant velocity until it returns to A, the journey lasting $t$ seconds, then by the clock that has remained at rest

the travelled clock on its arrival at $A$ will be $tv^2/2c^2$ seconds slow." Prove this statement. (*Note:* Elsewhere in his paper Einstein indicated that this result is an approximation, valid only for $v \ll c$.)

3. **Do you really want to do it?** You wish to make a round trip from earth in a spaceship, traveling at constant speed in a straight line for six months and then returning at the same constant speed. You wish further, on your return, to find the earth as it will be a thousand years in the future. (*a*) How fast must you travel? (*b*) Does it matter whether or not you travel in a straight line on your journey? If, for example, you traveled in a circle for one year, would you still find that a thousand years had elapsed by earth clocks when you returned?

4. **Synchronizing clocks.** Consider two clocks fixed along the $x$ axis of an inertial reference frame, one at $x = x_1$ and the other at $x = x_2$. In Section 2-1 we saw how to synchronize such clocks, using light signals. Here is another proposed method that, at first glance, may seem quite reasonable: Let a traveler move out along the $x$ axis with constant speed $v$, wearing a wristwatch. Let the traveler then set each of the two $x$ axis clocks to agree with the wristwatch as she passes them. What is wrong with this method of synchronization?

5. **Bob and Dave.** (*a*) In the spacetime diagram of Fig. B-2, how far apart are Bob and Dave when Bob turns around? (*b*) Suppose that Dave did not know beforehand when Bob was planning to turn around. When (by his own clocks and calendars) would Dave find out that Bob had done so? (*c*) If Bob's clock runs slow on the outbound trip (as it does), then why does it not run fast on the inbound trip, for which his velocity is reversed in sign?

6. **Bob changes his mind.** Suppose that Bob, after noting the passage of three years by his on-board clock, decides not to return to Dave but simply stops. He compares his on-board

clock with one of the local clocks belonging to the synchronized array of stationary clocks fixed in Dave's inertial frame. (*a*) What will this local clock read? (*b*) Draw Bob's world line for this new situation on a spacetime diagram.

7. **Bob is older than Dave this time.** Bob, once started on his outward journey from Dave, keeps on going at his original uniform speed of $0.8c$. Dave, knowing that Bob was planning to do this, decides, after waiting for three years, to catch up with Bob and to do so in three additional years. (*a*) To what speed must Dave accelerate to do so? (*b*) What will be the elapsed time by Bob's clock when they meet? (*c*) How far will they each have traveled when they meet, measured in Dave's original inertial reference frame? (*d*) Draw the world lines for Bob and Dave on a spacetime diagram and compare it with Fig. B-2. Notice that the present scenario is the mirror image of the one discussed in connection with that figure; there Dave turned out to be four years older than Bob when they reconvened; here Bob will turn out to be four years older than Dave [8].

8. **Bob and Dave are twins again**. Suppose that Bob and Dave each agree to follow the scenario described by Fig. B-2 for three years, each counting the years by his own onboard clock. Then Bob will come to rest and Dave will accelerate to $0.8c$ and eventually catch up with Bob. (*a*) What will be the total elapsed times on each of their clocks when they meet? (*b*) Draw a spacetime diagram and compare it carefully with that of Fig. B-2. Note the total symmetry of the situation. In the scenario as given originally, Dave turned out to be four years older than Bob when they met; in Problem 7, the reverse turned out to be true; in this case they turn out to be the same age at the end of their journeys [8].

9. **The twins talk it over.** Explain (in terms of heartbeats, physical and mental activities,

and so on) why the younger returning twin has not lived any longer than his own proper time even though his stay-at-home brother may say that he has. Hence, explain the remark: ''You age according to your own proper time.''

10. **The twin paradox and time dilation.** Time dilation is a symmetric (reciprocal) effect. The twin-paradox result is asymmetric (nonreciprocal). Discuss these two effects from this point of view and explain how they are related.

11. **Asymmetric aging and acceleration.** Is asymmetric aging always associated with acceleration? Can you have acceleration (of one or both twins) without asymmetric aging?

12. **Getting younger.** Can you think of any way to use space travel to reverse the aging process, that is, to get younger? Could you send your parents out on a long space voyage and have them be younger than you are when they get back?

## REFERENCES

1. Many articles on this topic are reproduced in Gerald Holton, Ed., *Special Relativity—Selected Reprints* (American Institute of Physics, New York, 1963). See also a number of Letters to the Editor in the September (1971) and the January (1972) issues of *Physics Today*.

2. C. G. DARWIN, ''The Clock Paradox in Relativity,'' *Nature,* **180,** 976 (1957).

3. O. R. FRISCH, ''Time and Relativity: Part I,'' *Contemp. Phys.,* **3,** 16 (1961), and O. R. Frisch, ''Time and Relativity: Part II,'' *Contemp. Phys.,* **3,** 194 (1962).

4. F. J. M. FARLEY, J. BAILEY, AND E. PICASSO, ''Experimental Verifications of the Theory of Relativity,'' *Nature,* **217,** 17 (1968). The test of time dilation using the muon storage ring—important in its own right—was an incidental feature of another investigation.

5. R. W. WILLIAMS, *Phys. Today* (January 1972), p. 11.

6. J. C. HAFELE and RICHARD E. KEATING, ''Around-the-World Atomic Clocks: Predicted Relativistic Time Gains,'' *Science,* **177,** 166 (1972), and J. C. Hafele and Richard E. Keating, ''Around-the-World Atomic Clocks: Observed Relativistic Time Gains,'' *Science,* **177,** 168 (1972).

7. See Edward M. Purcell, in Harry Woolf, Ed., *Some Strangeness in the Proportion: A Centennial Symposium to Celebrate the Achievements of Albert Einstein,* (Addison-Wesley, Reading, Mass. 1980), p. 106.

8. Problems 7 and 8 are adapted from Frank S. Crawford, ''Symmetrization of C. G. Darwin's Clock Paradox Scenario,'' *Am. J. Phys.* **51,** 1145 (1983).

# The Principle of Equivalence and the General Theory of Relativity

*I was sitting in a chair in the patent office at Bern when all of a sudden a thought occurred to me: "If a person falls freely he will not feel his own weight." I was startled. This simple thought made a deep impression on me. It impelled me toward a theory of gravitation.*

*Albert Einstein (1922)*

We have seen that special relativity requires us to modify the classical laws of motion. However, the classical laws of electromagnetism, including the Lorentz force law, remain valid relativistically. What about the gravitational force, that is, Newton's law of gravitation—does relativity require us to modify that? Despite its great success in harmonizing the experimental observations, Newton's theory of gravitation is suspect conceptually if for no other reason than that it is an action-at-a-distance theory. The gravitational force of interaction between bodies is assumed to be transmitted instantaneously, that is, with infinite speed, in contradiction to the relativistic requirement that the limiting speed of a signal is $c$, the velocity of light. And there are worrisome features about the interpretation of the masses in the law of gravitation. For one thing, there is the equality of inertial and gravitational mass, which in the classical theory is apparently an accident (see *Physics*, Part I, Sec. 16-4). Surely there must be some physical significance to this equality. For another thing, the relativistic concept of mass-energy suggests that even particles of zero rest mass will exhibit masslike properties (for example, inertia and weight). But such particles are excluded from the classical theory. If gravity acts on them, we must find how to incorporate this fact in a theory of gravitation.

In 1911 Einstein advanced his principle of equivalence, which became the starting point for a new theory of gravitation. In 1916, he published his theory of general relativity, in which gravitational effects propagate with the speed of light and the laws of physics are reformulated so as to be invariant with respect to accelerated (noninertial) observers. The equivalence principle is strongly confirmed by experiment. Let us examine this first.

## C-1 The Principle of Equivalence

Consider two reference frames: (1) a nonaccelerating (inertial) reference frame $S$ in which there is a uniform gravitational field and (2) a reference frame $S'$, which is accelerating uniformly with respect to the inertial frame but in which there is no gravitational field. Two such frames are physically equivalent. That is, experiments carried out under otherwise identical conditions in these two frames should give the same results. This is the *principle of equivalence*. If the principle is restricted to mechanical experiments alone, it may be taken as a consequence of Newtonian mechanics. Einstein, however, broadened the principle to include *all* physical experiments, including optical (that is, electromagnetic) experiments, and used this principle as the basis for his general theory of relativity.

To explore this principle further, let us imagine a spaceship to be at rest in an inertial reference frame $S$ in which there is a uniform gravitational field, say at the surface of the earth. Inside the spaceship, objects that are released will fall with an acceleration, **g,** in the gravitational field; an object that is at rest, such as an astronaut sitting on the floor, will experience a force opposing its weight. Now let the spaceship proceed to a region of outer space where there is no gravitational field. Its rockets are accelerating the spaceship, our new frame $S'$, with $\mathbf{a} = -\mathbf{g}$ with respect to the inertial frame $S$. In other words, the ship is accelerating away from the earth beyond the region where the earth's field (or any other gravitational field) is appreciable. The conditions in the spaceship will now be like those in the spaceship when it was at rest on the surface of the earth. Inside the ship an object released by the astronaut will accelerate downward relative to the spaceship with an acceleration **g.** And an object at rest relative to the spaceship, for instance, the astronaut sitting on the floor, will experience a force indistinguishable from that which balanced its weight before. From observations made in his own frame, the astronaut could not tell the difference between a situation in which his ship was accelerating relative to an inertial frame in a region having no gravitational field and a situation in which the spaceship was unaccelerated in an inertial frame in which a uniform gravitational field existed. The two situations are exactly equivalent.

Indeed, it follows that if a body is in a uniform gravitational field — such as an elevator in a building on earth — and is at the same time accelerating in the direction of the field with an acceleration whose magnitude equals that due to the field — such as the same elevator in free fall — then particles in such a body will behave as though they are in an inertial reference frame with no gravitational field. They will be free of acceleration unless a force is impressed on them. This is the situation inside an orbiting space laboratory, in which objects released by the astronaut will not fall relative to the orbiting vehicle (they appear to float in space) and the astronaut himself will be free of the force that countered the pull of gravity before launching (he feels weightless).

Einstein pointed out that, from the principle of equivalence, it follows that we cannot speak of the absolute acceleration of a reference frame, only a relative one, just as it followed from the special theory of relativity that we cannot speak of the absolute velocity of a reference frame, only a relative one. This analogy to special relativity is a formal one, for there is no absolute acceleration provided that we grant that there is also no absolute gravitational field. It also follows from the principle of equivalence (it is *not* an accident) that inertial mass and gravitational mass are equal (see Question 1).

# C-2 The Gravitational Red Shift

Now let us apply the principle of equivalence to see what gravitational effects there might be that are not accounted for in the classical theory. Consider a pulse of radiation (a photon) emitted by an atom $A$ at rest in frame $S$ (a spaceship at rest on the earth's surface, for example). A uniform gravitational field $\mathbf{g}$ is directed downward in $S$, the photon falling down a distance $d$ through this field before it is absorbed by the detector $D$ (see Fig. C-1a). To analyze what effect gravity has on the photon, let us consider the equivalent situation, shown in Fig. C-1b. Here we have an atom and a detector separated by a distance $d$ in a frame $S'$ in which there is no gravitational field, the frame $S'$ (a spaceship in outer space, for example) accelerating uniformly upward relative to $S$ with $a = g$. When the photon is emitted, the atom has some speed $u$ in frame $S$. The speed of the detector, when the photon reaches it, is $u + at$, where $t$ is the time of flight of the photon. But (see Question 6) $t$ is (approximately) $d/c$ and $a = g$, so the detector's speed on absorption is $u + g\, d/c$. In effect, the detector has an approach velocity relative to the emitter of $v = g\, d/c$, independent of $u$. This corresponds to a speed parameter $\beta(= v/c)$ of $g\, d/c^2$. Hence $v'$, the frequency received, is greater than $v$, the frequency emitted, the Doppler formula (see Eq. 2-31a) giving us

$$v' = v(1 + \beta + \tfrac{1}{2}\beta^2 + \cdots).$$

Because $g\, d/c^2 << 1$, we need keep only the first two terms in this equation and can write, for the fractional shift in frequency,

$$\frac{\Delta v}{v} = \frac{v' - v}{v} \cong \beta = g\,\frac{d}{c^2} . \tag{C-1}$$

By the principle of equivalence, we should obtain this same result in frame $S$. In this frame, however, $A$ and $D$ are at rest and there is no Doppler effect to explain the increase in frequency. There is a gravitational field, however, and the result in $S'$ suggests that this field might act on the photon. Let us explore this possibility by ascribing to the photon a *gravitational mass* equal to its inertial mass, $E/c^2$. Then,

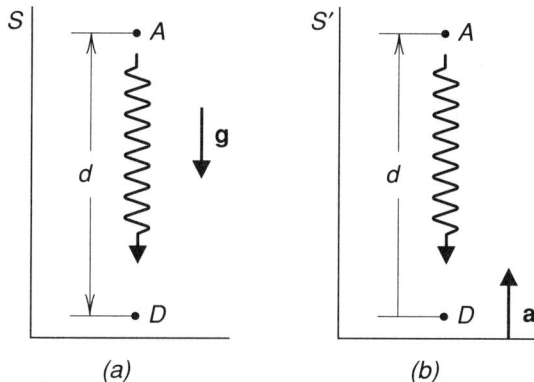

**FIGURE C-1.**   Gravitational red shift.

in falling a distance $d$ in a gravitational field of strength $g$, the photon gains energy $(E/c^2)g\,d$. How can we connect the energy $E$ to the frequency $\nu$? In the quantum theory, the connection is $E = h\nu$, *where $h$ is a constant called the Planck constant*. For the moment, let us use this relation so that the energy of the photon on absorption at $D$ is its initial emission energy plus the energy gained in falling from $A$ to $D$, or $h\nu + (h\nu/c^2)g\,d$. If we call this absorption energy $E' = h\nu'$, then we have

$$h\nu' = h\nu + \frac{h\nu g\,d}{c^2},$$

or, for the fractional frequency shift,

$$\frac{\Delta\nu}{\nu} = \frac{\nu' - \nu}{\nu} = g\,\frac{d}{c^2}, \qquad (\text{C-2})$$

the same result obtained in frame $S'$ (Eq. C-1).

Actually, it is not necessary to use quantum theory in deriving this result, a fact that we may suspect to be true because the Planck constant $h$, which is the characteristic constant of quantum theory, cancels out and does not appear in Eq. C-2. We can show in relativity itself that $E$ is proportional to $\nu$, because it follows, from the relativistic transformation of energy and momentum, that the energy in an electromagnetic pulse changes by the same factor as its frequency when observed in a different reference frame. The conclusion then is that, in falling through a gravitational field, light gains energy and frequency (its wavelength decreases and we say that it is shifted toward the blue). Clearly, had we reversed emitter and detector, we would have concluded that in rising against a gravitational field light loses energy and frequency (its wavelength increases and we say that it is shifted toward the red).

Even with $d$ in Eq. C-2 being the distance from sea level to the top of the highest mountain, the predicted value of $\Delta\nu/\nu$, the fractional frequency shift, is only about $10^{-12}$. Nevertheless, Pound and Rebka [1] in 1960 were able to confirm the prediction, using the 74-ft-high Jefferson tower at Harvard! For this small distance we have

$$\frac{\Delta\nu}{\nu} = \frac{g\,d}{c^2} = \frac{(9.8 \text{ m/s}^2)(22.5 \text{ m})}{(3.0 \times 10^8 \text{ m/s})^2} = 2.5 \times 10^{-15},$$

an incredibly small effect. By using the Mössbauer effect (which permits a highly sensitive measurement of frequency shifts) with a gamma-ray source, and taking admirable care to control the competing variables, Pound and Rebka observed this gravitational effect on gamma-ray photons and confirmed the quantitative prediction with a precision of about 10 percent. In a subsequent refinement of the original experiment, Pound and Snider [2] in 1965 found, comparing experimental observation with theoretical prediction, that

$$\frac{(\Delta\nu/\nu)_{\text{exp}}}{(\Delta\nu/\nu)_{\text{theory}}} = 0.9990 \pm 0.0076.$$

The most precise verification of the gravitational frequency shift, however, was reported in 1980 by Vessot and his co-workers [3]. They fired a space probe vertically upward to a height of 10,000 km, the probe containing a microwave transmitter whose frequency was accurately controlled by a hydrogen maser. By comparing the frequency received at the ground station with that of a similar ground-based maser, these workers were able to measure, among other quantities, the gravitational frequency shift of the space-borne "clock." They concluded that the predictions of relativity theory of the effect of changes in gravitational potential on this clock during this experiment can be relied on to a precision of one part in 14,000.

We can easily generalize our result (Eqs. C-1 and C-2) to photons emitted from the surface of stars and observed on earth. Here we assume that the gravitational field need not be uniform and that the result depends only on the difference in gravitational potential between the source and the observer. Then, in place of $g\,d$ we have $GM_s/R_s$, where $M_s$ is the mass of the star of radius $R_s$, and because the photon *loses* energy in rising through the gravitational field of the star, we obtain

$$\frac{\Delta\nu}{\nu} = -\frac{GM_s}{R_sc^2}.$$ (C-3)

This effect is known as the *gravitational red shift*, for light in the visible part of the spectrum will be shifted in frequency toward the red end. This effect is distinct from the Doppler red shift from receding stars. Indeed, because the Doppler shift is much larger, the gravitational red shift is difficult to verify, in part because the masses and radii of stars other than the sun are not well known. The effect has been confirmed with certainty, however, for the white dwarf star Sirius B and verified to a precision of about 5 percent for light from the sun.

# C-3 General Relativity Theory

**Features and Foundations**  A full treatment of the general theory of relativity is beyond the scope of a book at this level [4]. We therefore limit ourselves to a qualitative description of some of its important characteristics. Let us return, then, to Fig. C-1a, in which detector $D$ measures a greater frequency than source $A$ emits. It may appear strange that a frequency can increase with no relative motion of source and detector. After all, $D$ surely receives the same *number* of vibrations that $A$ sent out in its light pulse (photon). Even the distance of separation remains constant between $A$ and $D$, so how do we interpret the measured frequency increase? The answer, once again, is that there is a disagreement about time. That is, frequency is the number of vibrations per unit time, so the frequency difference must be due to a time difference; the rate of $A$'s clock must differ from the rate of $D$'s clock.

We conclude that clocks in a region of higher gravitational potential (that is, higher above the earth's surface) run faster than those in a region of lower gravitational potential. Let the emitting atom $A$ be the clock, for example, its rate being the frequency of its emitted radiation. Then the higher up in the gravitational field the atom is, the higher its frequency appears to be to $D$ (that is, compared to the same atom radiating at $D$). Similarly, if we reverse $A$ and $D$, then the lower down in the gravitational field the atom is, the lower its frequency appears to be to $D$ (that is, compared to the same atom radiating at $D$).

We can use these considerations to throw some light on the twin paradox of Supplementary Topic B. Consider the experiment (see pages 81 and 163) in which a "clock" is placed on the rim of a rotor, a radial distance $R$ from the axis. The central fact is that this clock (the traveling twin) runs slow compared to a clock located on the rotor axis (the stay-at-home twin). In our earlier discussion we took the point of view of an (inertial) observer on the rotor axis and explained the phenomenon as the familiar time dilation of special relativity, associated with the speed $v$ of the traveling clock. We can equally well take the point of view of a (noninertial) observer on the rotor rim. This observer, using an argument from general relativity, would explain the same fact by asserting that the clock is in an effective gravitational field of magnitude $v^2/R$ that acts on the clock and slows it. Both approaches yield the same numerical result for the rate difference between the axis clock and the rim clock, as indeed the principle of relativity requires.

All of the above conclusions about the effect of a gravitational field on the rate of a clock follow from the principle of equivalence, first enunciated by Einstein. It was also Einstein who called attention to the gravitational red shift required from theory and to the need to ascribe a gravitational mass $m$ ($= E/c^2$) to an energy $E$. Still another of his results was that the direction of the velocity of light is not constant in a gravitational field; indeed, light rays *bend* in such a field because of their gravitational mass, and Einstein predicted that this bending would cause a displacement in the apparent positions of fixed stars that are seen near the edge of the sun.

When we discussed the special theory of relativity in Chapters 1, 2, and 3, we deliberately excluded gravitational fields from consideration, without even mentioning that we were doing so; thus the possibility that such fields might affect clock rates never arose. Their presence would certainly have complicated, for example, our discussion of simultanity (see Section 2-1) on which our derivation of the Lorentz transformation equations (Section 2-2) was based. Thus it becomes clear that a more general theory* is needed, which takes into account the principle of equivalence and which generalizes even that principle to include nonuniform gravitational fields as well. Furthermore, gravitational effects themselves must be treated by a theory in which the propagation speed is finite. In a series of papers [5], Einstein formulated just such a general theory of relativity.

We have seen that the special theory of relativity asserts that physics is the same for all *inertial* observers, assuming only that gravitational forces are excluded. The general theory goes further and asserts that physics is the same for *all* observers, whether or not they are in inertial (unaccelerated) reference frames and whether or not gravitational forces are present. Thus the special theory appears as a limiting case of the general theory.

Einstein was lead to the general theory by the realization of the deep significance of the principle of equivalence. By use of this principle it is possible to "transform

---

*Just as Galilean relativity is a limiting case of special relativity, so special relativity itself is a limiting case of general relativity. The gravitational field near the earth is so weak that there are no readily measurable discrepancies between special and general relativity. We usually operate in the special relativistic limit of general relativity.

away''—as the technical phrase goes—inhomogeneous gravitational fields by having at *each* point a *different* accelerated reference frame that replaces the local (infinitesimal) gravitational field there. In such local frames, the special theory of relativity *is* valid, so the invariance of the laws of physics under a Lorentz transformation applies to infinitesimal regions. Second, through an invariant spacetime metric that follows from this, we can link geometry to gravitation and geometry becomes non-Euclidean. That is, the presence of a large body of matter causes spacetime to warp in the region near it so that spacetime becomes non-Euclidean. This warping is equivalent to the gravitational field. The curvature of spacetime in general relativity replaces the gravitational field of classical theory. Hence, the geometry of spacetime is determined by the presence of matter. In this sense, geometry becomes a branch of physics. The fact that special relativity is valid in small regions corresponds to the fact that Euclidean geometry is valid over small parts of a curved surface. In large regions, special relativity and Euclidean geometry need not apply, so the world lines of light rays and of inertial motion need not be straight; instead, they are geodetic, that is, as straight as possible.

The relationship between matter and geometry has been aptly summarized by the statement: ''Geometry tells matter how to move, and matter tells geometry how to curve.'' General relativity has, in fact, been described as *geometrodynamics,* to reflect this interaction [6]. Figure C-2 suggests the interaction between matter and geometry by what has been called the rubber sheet analogy. In Fig. C-2*a* a stretched rubber sheet is shown, marked with coordinate lines. No matter is present, the geometry is Euclidean, special relativity holds throughout, and particles not acted on by forces move in straight lines as suggested by the trajectory shown. In Fig. C-2*b,* however, a heavy ball bearing has been placed on the sheet, deforming it and its coordinate lines. Now matter *is* present and is affecting the geometry, which becomes non-Euclidean; general relativity must be used, and the trajectory shown suggests the motion of a particle deflected by a centrally directed gravitational force.

**Experimental Tests**  The gravitational red shift, discussed in the previous section, is a test of the equivalence principle, as formulated by Einstein, rather than a specific test of the theory itself. A second critical test of this principle is the verification of the observation that, in free fall, all bodies fall with the same acceleration, regardless of their mass or composition. This is equivalent to the assertion that the gravitational mass of a body is equal to its inertial mass. Measurement has confirmed that these quantities are indeed equal, to within one part in $10^{12}$. Thus the foundations of the general theory of relativity seem to be firmly rooted in experiment. Note that Newton's theory of gravity has no explanation at all for the gravitational red shift and can only view the equality of gravitational and inertial mass as a coincidence of astonishing proportions.

As for specific predictions of the theory itself, there are three that have been subject to careful experimental scrutiny. (1) The precession of the perhelion of the planet Mercury (Fig. C-3*a*) should differ from the classical prediction, the difference being about 43 seconds of arc per century. (2) The positions of stars whose light passes near the edge of the sun (observed, say, during a total eclipse) should

*(a)*

*(b)*

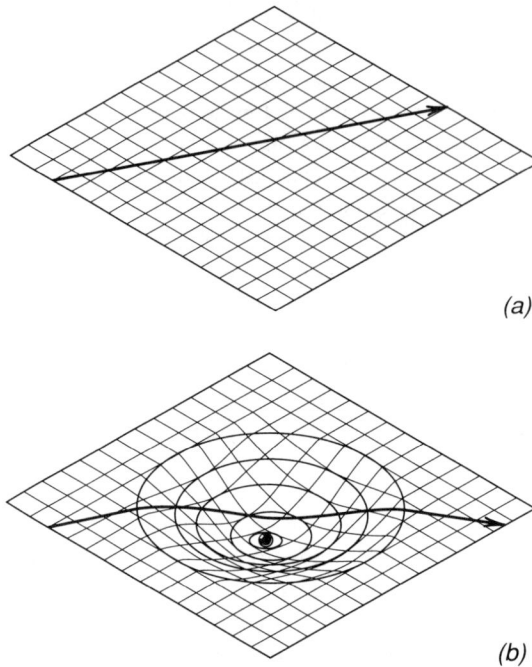

**FIGURE C-2.**   The rubber sheet analogy for the interaction between matter and geometry.

be displaced (due to deflection in the sun's gravitational field) by 1.75 seconds of arc from their positions observed at night (Fig. C-3*b*). (3) There should be a time delay in radio or radar signals (sent, say, from earth to one of the other planets and back again) if the light path passes near the sun (Fig. C-3*c*). We consider each prediction in turn.

1. The observed rate of precession of the perhelion of Mercury about the sun is 5599.74 ± 0.41 seconds of arc per century. Of this, all but 43.11 ± 0.45 seconds of arc per century can be accounted for by the fact that the earth is not an inertial reference frame and by the Newtonian gravitational effects of other planets. The prediction of the general theory of relativity is 43.03 seconds of arc per century, so the ratio of the observed to the predicted value is 1.00 ± 0.01, a striking confirmation of the theory. Perhelion shift rates have also been measured for other planets, and for the asteroid Icarus; the agreement is again within the error of the measurements, but the precision of those measurements is not nearly as good as it is for the planet Mercury.

2. The bending of starlight as the rays pass near the rim of the sun has been measured many times during solar eclipses. The angular deflection predicted by theory is 1.75 arc seconds. The observed deflections, when the possibility of systematic errors is taken into account, are judged to agree with this to a precision within the range of 10 to 20 percent. More recently, radio emissions from distant quasars, passing near the sun's rim on their path to earth, have

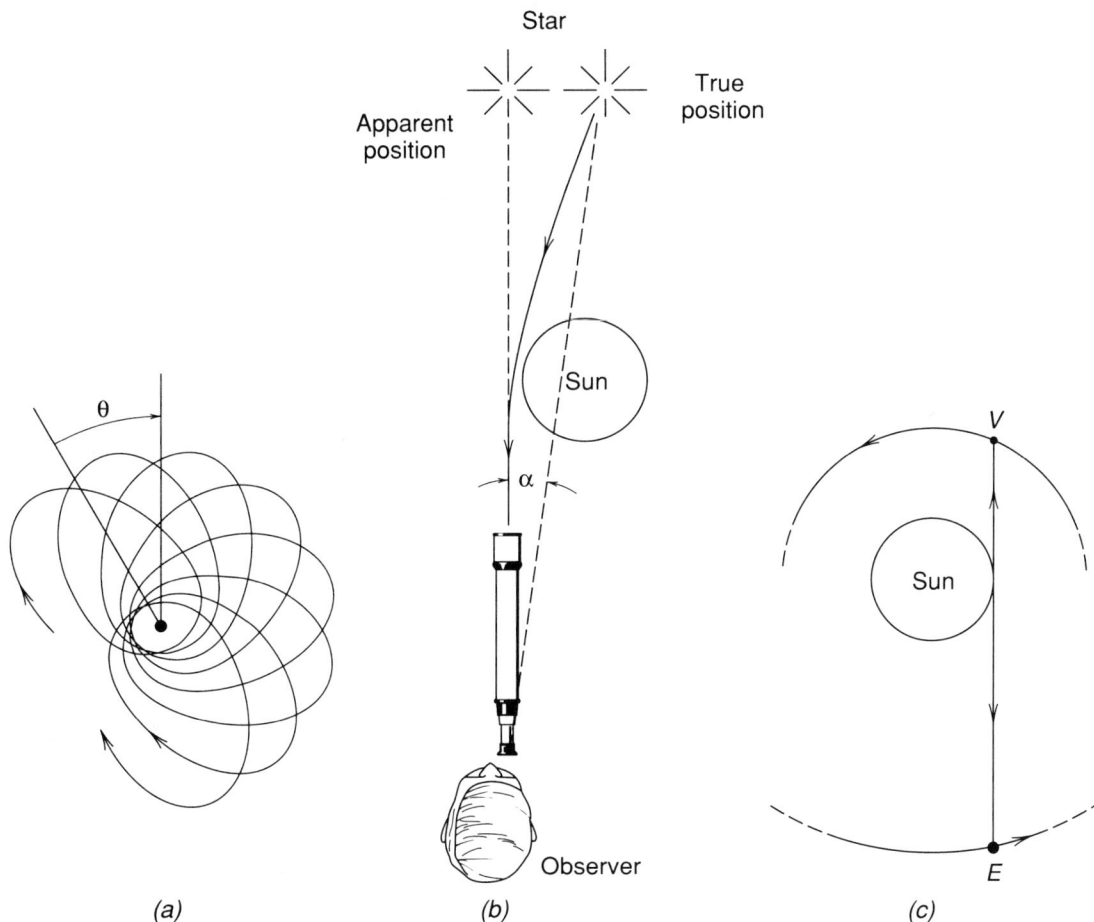

**FIGURE C-3.** *Three solar system tests of general relativity.* (a) A planet moving in a precessing elliptical orbit about the sun. The perhelion, or point of closest approach, shifts by an angle θ after each excursion. (b) Light passing the sun from a star to earth is deflected, making the star appear displaced away from the sun by an angle (= 1.75 arc seconds). (c) A radar pulse sent from earth to Venus and back is delayed slightly if its path passes near the sun. All drawings are schematic and clearly not to scale.

been observed to be deflected by amounts that agree with the prediction of the general theory to within 1 percent or so. Fomalont and Sramek, for example, measured a solar-rim deflection of $1.775 \pm 0.019$ arc seconds by this method [7].

3. The delay in the echo times of radio signals reflected from Venus, when that planet is on the far side of the sun from us, have been measured by Shapiro and his co-workers [8] and found to agree with the predictions of the general theory to within about 2 percent.

See Ref. 6 (part IX; see Summary on p. 1129) for a detailed analysis of the experimental tests to which Einstein's general theory has been subjected as of 1972. These authors, commenting on the success of Einstein's theory with respect to competing theories that have been advanced from time to time, conclude: "As experiment after experiment has been performed, and one theory of gravity after another has fallen by the wayside a victim of the observations, Einstein's theory has stood firm. No purported inconsistency between Einstein's laws of gravity has ever surmounted the test of time." (See also Ref. 11.)

**Predictions and Consequences** There are several other areas now under active investigation that relate to Einstein's general theory of relativity:

1. *Gravitational lenses.* In 1979 two "twin quasars" were discovered, after intensive investigation at both optical and radio wavelengths and after much theoretical analysis, not to be two distinct objects at all but rather to be images of a single distant quasar, formed by the bending of light rays by the gravitational action of an intervening galaxy [9]. Since that time several other gravitationally lensed quasar images have been found. See Problem 15.

2. *Black holes.* The term originates from the possibility that a body of a given size might be massive enough for its strong gravitational field to prevent the escape of light from it. In general relativity, theory predicts that, under certain conditions, a sufficiently massive body can indeed undergo gravitational collapse to form a black hole. Such an object can be detected only by secondary effects associated with its strong gravitational field. Thus, if a black hole forms a binary system with an ordinary star, it is possible for matter to be transferred from the star to the black hole. As the matter spirals in toward the back hole, the particles of matter collide with each other and may generate sufficient heat that they emit x-rays. On this basis, many astrophysicists believe that one component of the x-ray binary identified as Cygnus X-1 is a black hole. Others believe that there is strong evidence for the existence of a black hole at the center of our own galaxy.

3. *Gravity waves.* Just as an accelerated charge emits electromagnetic radiation traveling at the speed of light, so implicit in the general theory of relativity is the prediction that an accelerated mass can emit gravitational waves, also traveling at the speed of light. Much effort is now being expended to detect such radiation unambiguously. The best evidence for the existence of such waves, however, is indirect but nevertheless convincing. The binary pulsar identified as PSR 1913 + 16 is thought to emit gravitational waves, because of the acceleration associated with the orbital motion of the pulsar and its companion star about their center of mass [10]. The energy presumably carried away by gravitational radiation results in a slow decrease in the period of this orbital motion, amounting to 67 ns per orbit. Fortunately, this slow decrease has measurable effects on the arrival times of the pulses from the pulsar. In particular, the times of periastron (that is, the times when the components of the binary are closest together in their orbital motions) can be measured with some precision. The general theory of relativity predicts that loss of energy by gravitational radiation will cause the periastron times to be cumulatively displaced—as time goes on—when compared with the periastron times expected for a hypothetical system that does not radiate. Figure C-4 shows the excellent agreement between theory (solid curve) and observation.

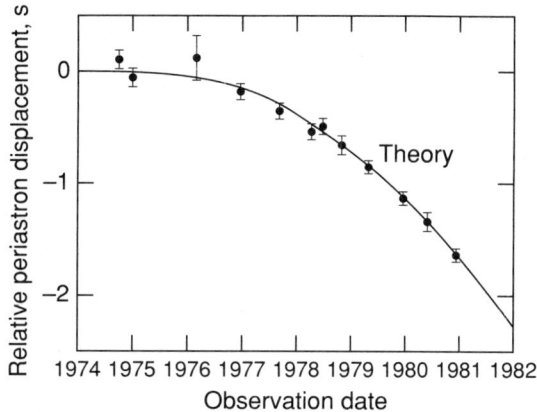

**FIGURE C-4.** The points show the deviations in time of periastron passage for the binary pulsar PSR 1913 + 16 compared to a hypothetical binary that does not emit gravity radiation. The solid curve is the prediction of the general theory of relativity [10].

4. *Cosmological predictions*. The general theory of relativity is one of the greatest intellectual achievements of all time. Its originality and unorthodox approach exceed that of special relativity. And far more than special relativity, it was almost completely the work of a single man, Albert Einstein. The philosophic impact of relativity theory on the thinking of man has been profound and the vistas of science opened by it are literally endless. To quote Max Born, writing in 1962: "The idea first expressed by Ernst Mach, that the inertial forces are due to the total system of the fixed stars, suggests the application of the theory of general relativity to the whole universe. This step was actually made by Einstein in 1917, and from that time dates the modern development of cosmology and cosmogony, the sciences of the structure and genesis of the cosmos. This development is still in full swing and rich in important results, though far from final conclusions." In 1972 Wheeler wrote: "Einstein's description of gravitation as curvature of spacetime led directly to that greatest of all predictions of his theory, that the universe itself is dynamic. Physics still has far to go to come to terms with this amazing fact and what it means for man and his relation with the universe."

## QUESTIONS AND PROBLEMS

1. **Free fall, the same for all.** Starting from the fact that all bodies that are free of forces move with uniform velocity relative to an inertial reference frame, then considering the motion of those bodies in an accelerated frame, and finally using the principle of equivalence, show that all bodies fall with the same acceleration in a uniform gravitational field. Hence the equality of gravitational and inertial mass.

2. **Transforming away gravity.** Can gravity be regarded as a "fictitious" force, arising from the acceleration of one's reference frame relative to an inertial reference frame, rather than a "real" force? Discuss an analogy with the forces encountered on a rotating (that is, accelerating) reference frame such as a merry-go-round.

3. **The equivalence principle and the bending of light.** (a) Consider the path of a light ray to be straight in an unaccelerated reference frame where there is no gravitational field. What will the path look like if the frame accelerates upward with an acceleration **a**? (b) Use the principle of equivalence to show that we should expect the path to look just the same if a homogeneous gravitational field **g** (= −**a**) is present.

4. **The gravitational mass of light.** A photon may be viewed as a particle of light, its energy being given by $hv$, where $v$ is the frequency and $h$ (= $6.63 \times 10^{-34}$ J · s) is the Planck constant. In terms of $m_e$, the electron rest mass, what is the gravitational mass of a 600-nm visible-light photon?

5. **Rest mass or total mass?** The mass of a moving particle can be considered to be either its rest mass or its total relativistic mass. Which mass is appropriate for gravitation, that is, for the concept of mass as causing gravitation or being responsive to gravitation? Consider the special case of light, which has zero rest mass.

6. **Clearing up a point.** Why is the relation $t = d/c$, used for the light pulse's (that is, the photon's) time of flight in frame $S'$ of Fig. C-1b, only approximate, rather than exact?

★7. **The clock on the top floor is running fast.** An atomic clock is placed in the basement of the World Trade Center in New York, a second similar clock being placed on the 110th floor, 1,350 ft higher. How long will the clocks have to run before (because of the difference in gravitational potential be-

tween their locations) the upper clock gains 1.0 μs by comparison with the lower clock? Before the upper clock gains 1.0 ns?

8. **Comparing clocks.** Occasionally, an atomic clock is flown from Washington, D.C. to Boulder, Colorado, a distance of 2,400 km, for intercomparison with standard clocks kept at the Boulder Laboratory of the National Bureau of Standards. (a) How much time has been lost by the transported clock upon its arrival in Boulder, by virtue of its trip from Washington at an assumed average speed of 650 mi/h? Note that this effect can be accounted for by special relativity alone; see Supplementary Topic B. (b) If the transported clock stays in Boulder for ten days before being returned to Washington, what correction must be subtracted from its reading, by virtue of the fact that Boulder is 1,600 m higher than Washington? Note that this effect is a general-relativistic effect.

9. **The gravitational red shift for light from a white dwarf star.** (a) Show that, in terms of wavelength, the gravitational red shift (see Eq. C-3) can be written as

$$\lambda = \lambda_0 \left( 1 + \frac{GM_s}{R_s c^2} \right).$$

(b) Light of wavelength 600.0 nm is emitted by a white dwarf star. Due to the gravitational red shift, what is the wavelength received on earth? (c) What relative speed between the white dwarf and the earth would produce this same wavelength shift? Assume that the white dwarf has a mass (= $2.0 \times 10^{30}$ kg) equal to that of the sun but a radius (= $8.0 \times 10^6$ m) only about 1.1 percent as great as the solar radius.

10. **Gravitational red shift for sunlight.** The solar spectrum contains the sodium D line, which is a doublet, one of its components having a wavelength of 588.997 nm. Calculate the wavelength shift expected for this component because of (a) the gravitational red shift and (b) the Doppler effect associated with the sun's rotation, assuming in this

latter case that the light originates at the rim of the sun at its equator. The sun's mass and radius are $2.0 \times 10^{30}$ kg and $7.0 \times 10^8$ m, respectively. The sun's period of rotation, measured at the equator, is 26 d. (*Hint:* Use the result of Problem 9a.)

**11. A gravitational blue shift.**   A satellite in orbit at an altitude of 200 km sends a radio signal of frequency 1020 MHz to the ground station directly below on earth. What is the frequency shift of the signal, as received by the ground station, associated with the earth's gravitational field?

**12. Deflection of light à la Newton.**   It can be shown by classical mechanics that the deflection $\varphi$ of a particle of initial speed $v_0$ whose extended initial trajectory passes within a distance $b$ (the so-called impact parameter) of a body of mass $M$ is given by (see Fig. C-5)

$$\tan \frac{\varphi}{2} = \frac{GM}{bv_0^2}.$$

In deriving this expression it has been assumed that $m << M$, where $m$ is the mass of the deflected particle; note that $m$ does not

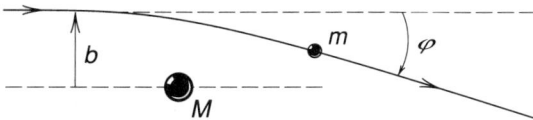

**FIGURE C-5.**   Problem 12.

appear in the above expression. Let us now apply this purely Newtonian result to a "particle" of light whose extended trajectory grazes the rim of the sun. Putting $v_0 = c$, $b = R_s$, and $M = M_s$ in the above expression, and assuming further that, because $\varphi$ is small, $\tan \varphi/2 \cong \varphi/2$, leads to

$$\varphi = \frac{2GM_s}{R_s c^2}.$$

Evaluate this quantity and show that the classically predicted value is just half of the value predicted by general relativity (= 1.75 seconds) for this quantity.

**13. How big is a black hole?**   If an object of mass $M$ has a radius less than a certain critical value, it will undergo gravitational collapse and become a black hole. This critical radius—known as the *Schwarzschild radius*—is given

$$R_s = \frac{2GM}{c^2}.$$

(*a*) "Derive" this result from classical mechanics in this way: Assume (incorrectly) that the kinetic energy of light is $\frac{1}{2} mc^2$ and (also incorrectly) use Newton's law of gravitation to find the radius of a mass $M$ from which the escape velocity is $c$. (These two errors happen to cancel, giving a correct result!) (*b*) Calculate the Schwarzschild radius for an object having a mass equal to that of the earth (= $6 \times 10^{24}$ kg), of the sun (= $2 \times 10^{30}$ kg), and of our Milky Way galaxy (≅ $3 \times 10^{41}$ kg).

**14. The density of black holes.**   (*a*) Show that the average density $\rho$ of a black hole, that is, of the material inside the Schwarzschild radius (see Problem 13), is given by

$$\rho = \frac{3c^6}{32\pi G^3 M^2}.$$

(*b*) Evaluate this density numerically for the three objects described in Problem 13b. (*c*) Note that, if $M$ is large enough, the average density of a black hole can be quite low. For what mass $M$ would the average density be equal to that of water?

**15. A gravitational lens.**   Mass $M$ in Fig. C-6 (a galaxy) is along the line of sight from earth $E$ to a distant quasar $Q$. It deflects the light from the quasar so that an earth observer sees two images $A$ and $B$ of the quasar, each displaced from the quasar's actual position. The angle $\alpha$ by which each light ray is bent is, from the general theory of relativity,

$$\alpha = \frac{4GM}{bc^2},$$

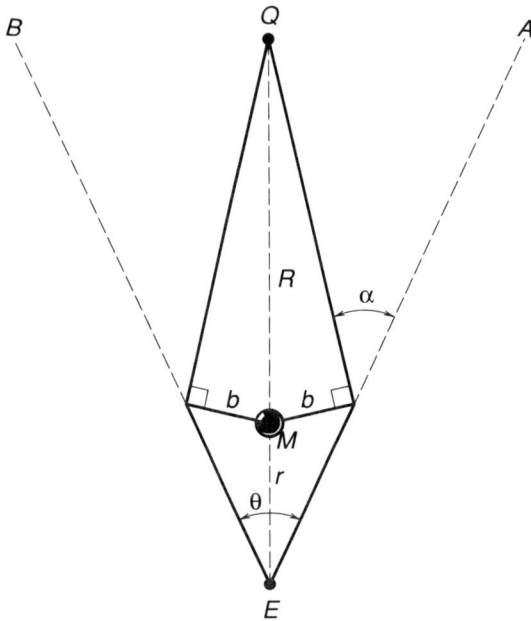

**FIGURE C-6.** Problem 15.

where $b$ is the so-called impact parameter of the incident light ray. The twin quasars identified as $0957 + 561$, at a distance $R$ from earth of about $6 \times 10^{24}$ m, appear in the sky in virtually the same direction as a galaxy whose distance $r$ from earth is about $2 \times 10^{24}$ m. Assume that the mass of this galaxy is about the same as that of the Milky Way galaxy ($= 3 \times 10^{41}$ kg). Assuming that the quasars are two images of a single quasar, calculate the expected angular separation $\theta$ between the images. (*Note:* the angles $\alpha$ and $\theta$ are very small; the calculated answer agrees well with observation.)

★**16. Precession of the perhelion.** According to the general theory of relativity, the angular advance per revolution of a planet's orbit is given by

$$\theta = \frac{A}{r},$$

where $A$ is a constant, the same for all planets, and $r$ is the mean radius of the planet's orbit. $P_M$, the rate of precession of the perhelion of the orbit of Mercury due to general relativistic effects, is 43.11 arc seconds/century. What is $P_E$, the corresponding rate of precession of the perhelion of the earth's orbit? The ratio of the mean radius of Mercury's orbit to that of the earth is 0.386. (*Hint:* Use Kepler's third law.)

**17. Gravitational waves versus electromagnetic waves.** What similarities might you expect in the production, detection, and nature of gravitational and electromagnetic waves?

**18. Life on the rim of a centrifuge.** A clock is located in an inertial frame on the axis of a centrifuge. A similar clock is located on the centrifuge rim, a radial distance $R$ from the axis. Show that the fractional difference in the rates of the clocks is given, in the case of $v \ll c$, by $v^2/2c^2$, in which $v$ is the speed of the rim clock. Derive this result from the point of view of (*a*) an axis observer, using special relativity, and also (*b*) a rim observer, using the equivalence principle. [*Hint:* Replace the centripetal acceleration by an equivalent gravitational field. The fractional difference in clock rates is given by $(V_r - V_a)/c^2$, where the quantity in parentheses is the difference in gravitational potential between the two clock sites.]

**19. The equivalence principle in Fantasy Land.** Figure C-7 suggests the design of a spaceship that uses the principle of equivalence to permit high accelerations while subjecting passengers in the life capsule to a standard acceleration of 1 "gee". $M$ is a sphere whose radius is 12.0 m, constructed of "compressed matter, electromagnetically stabilized" to a density of $1.20 \times 10^{12}$ kg/m³. The position of the life capsule along the central shaft can be adjusted in flight. (*a*) If the ship is coasting in free space at constant velocity, what must be the separation $h$ between the center of the life capsule and the center of sphere $M$ if the passengers are to experience a 1.0-gee environment? (b) At an acceleration of 35 gee for the spaceship as a whole, what must this separation be to maintain the 1.0-gee environment? What is the so-called tidal force (measured in gees per meter, along the central shaft) in the life capsule under these conditions? (Adapted from Charles Sheffield, ''Moment of Inertia,'' *Analog Science Fiction/Science Fact,* October 1980. In the story the life capsule gets stuck in position on the central shaft while the spaceship is in high-gee acceleration with the consequence

that, if the acceleration is reduced, the passengers will be flattened by the gravitational forces due to sphere $M$. Read the story to see how they escape, again using the equivalence principle.)

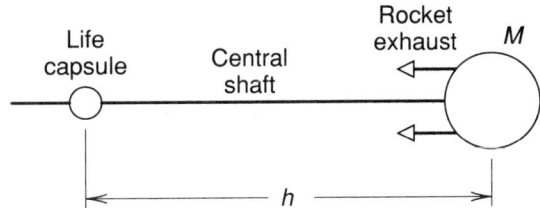

**FIGURE C-7.**    Problem 19.

# REFERENCES

1. R. V. Pound and G. A. Rebka, ''Apparent Weight of Photons,'' *Phys. Rev. Lett.,* **4,** 337 (1960).

2. R. V. Pound and J. L. Snider, ''Effect of Gravity on Gamma Radiation,'' *Phys. Rev. B,* **140,** 788 (1965).

3. R. F. C. Vessot, M. W. Levine, E. M. Mattison, E. L. Blomberg, T. E. Hoffman, G. U. Nystrom, and B. F. Farrel; R. Decher, P. B. Eby, C. R. Baugher, J. W. Watts, D. L. Teuber, and F. D. Wills, ''Test of Relativistic Gravitation with a Space-Borne Hydrogen Maser,'' *Phys. Rev. Lett.,* **45,** 2081 (1980).

4. Max Born, *Einstein's Theory of Relativity* (Dover, New York, 1965), chap. VII; Albert Einstein, *Relativity, The Special and the General Theory* (Crown, New York, 1952).

5. A. Einstein, H. Minkowski, H. A. Lorentz, and H. Weyl, *The Principle of Relativity: A Collection of Original Memoirs on the Special and General Theory of Relativity,* notes by A. Sommerfeld (Dover, New York, 1952).

6. Charles W. Misner, Kip S. Thorne, and John Archibald Wheeler, *Gravitation* (W. H. Freeman, San Francisco, 1973), chap. 1.

7. E. B. Fomalont and R. A. Sramek, ''A Confirmation of Einstein's General Theory of Relativity by Measuring the Bending of Microwave Radiation in the Gravitational Field of the Sun,'' *Astrophys. J.,* **199,** 749 (1975).

8. Irwin I. Shapiro, Michael E. Ash, Richard P. Ingalls, and William B. Smith; Donald B. Campbell, Rolf B. Dyce, Raymond F. Jurgens, and Gordon H. Pettingill, ''Fourth Test of General Relativity: New Radar Result,'' *Phys. Rev. Lett.,* **26,** 1132 (1971).

9. Frederic H. Chaffee, Jr., ''The Discovery of a Gravitational Lens,'' *Sci. Am.* (November 1980).

10. Joel M. Weisberg, Joseph H. Taylor, and Lee A. Fowler, ''Gravitational Waves from an Orbiting Pulsar,'' *Sci. Am.* (October 1981).

11. M. Will Clifford, *Was Einstein Right?* (Basic Books, New York, 1986).

# Answers to Problems

*It is better to know some of the questions than all of the answers.*

*James Thurber (1894-1961)*

## CHAPTER 1

1. (*a*) $2.59 \times 10^{13}$.   (*b*) 0.30.   (*c*) 3.26.
   (*d*) $3.33 \times 10^{-24}$.
2. (*a*) $3 \times 10^{-18}$.   (*b*) $2 \times 10^{-12}$.   (*c*) 8.2 $\times 10^{-8}$.
   (*d*) $6.4 \times 10^{-6}$.   (*e*) $1.1 \times 10^{-6}$.
   (*f*) $3.7 \times 10^{-5}$.   (*g*) $9.9 \times 10^{-5}$.   (*h*) 0.10.
3. (*a*) 30.0 cm/ns.   (*b*) 984 ft/μs.   (*c*) 1 ly/y.
   (*d*) $3.00 \times 10^{8} \ \text{J}^{1/2} \cdot \text{kg}^{-1/2}$.
   (*e*) $30.5 \ \text{MeV}^{1/2} \cdot \text{u}^{-1/2}$.
   (*f*) $3.00 \times 10^{8} \text{m} \cdot \text{F}^{-1/2} \cdot \text{H}^{-1/2}$.
4. (*a*) Light beam in a vacuum, electron beam, light beam in air.   (*b*) $1.1 \times 10^{-14}$ s.

(*c*) $9.7 \times 10^{-9}$ s.
5. (*a*) $3.37 \times 10^{-2}$ m/s$^2$.   (*b*) $5.94 \times 10^{-3}$ m/s$^2$.
   (*c*) $2.91 \times 10^{-10}$ m/s$^2$.
6. (*a*) $6.7 \times 10^{-10}$ s.   (*b*) $2.2 \times 10^{-18}$ m.
12. (*a*) $W_T = 48.0$ J; $W_G = 293$ J.   (*c*) For $T$: $K_i = 0$ and $K_f = 48.0$ J; for $G$: $K_i = 313$ J and $K_f = 606$ J.
13. (*b*) 24.0 J.
14. See box 14 below.
15. See box 15 below.

14.

|  | Symbol | Steven's Data | Sally's Data |
|---|---|---|---|
| Particle masses, kg | $m_1$ | 0.107 | 0.107 |
|  | $m_2$ | 0.345 | 0.345 |
| Particle velocities, m/s | $v_1$ | +3.25 | +0.75 |
|  | $v_2$ | 0.00 | −2.50 |
|  | $v'_1$ | −1.71 | −4.21 |
|  | $v'_2$ | +1.54 | −0.96 |
| Total system momentum, kg·m/s | $P$ | 0.348 | −0.782 |
|  | $P'$ | 0.348 | −0.782 |
|  | $P' - P$ | 0.000 | 0.000 |
| Total system kinetic energy, J | $K$ | 0.565 | 1.11 |
|  | $K'$ | 0.565 | 1.11 |
|  | $K' - K$ | 0.000 | 0.000 |

15.

|  | Symbol | Steven's Data | Sally's Data |
|---|---|---|---|
| Particle masses, kg | $m_1$ | 0.107 | 0.107 |
|  | $m_2$ | 0.345 | 0.345 |
| Particle velocities, m/s | $v_1$ | +3.75 | +0.75 |
|  | $v_2$ | 0.00 | −2.50 |
|  | $v'_1$ | −0.78 | −3.28 |
|  | $v'_2$ | +1.25 | −1.25 |
| Total system momentum, kg·m/s | $P$ | 0.348 | −0.782 |
|  | $P'$ | 0.348 | −0.782 |
|  | $P' - P$ | 0.000 | 0.000 |
| Total system kinetic energy, J | $K$ | 0.565 | 1.108 |
|  | $K'$ | 0.302 | 0.845 |
|  | $K' - K$ | −0.263 | −0.263 |

**17.** (b) 0.81%; 0.073%.   (c) 0.31%; 0.024%.

**20.** (a) 8.6° into the wind.   (b) Cross-wind, by 18 s.

**21.** 140.

**22.** 6.4 cm.

**25.** (a) After ~2 s a ring of light appears, its plane being at right angles to the direction of the ether wind. The ring splits into two, the two separate rings moving away from each other, shrinking, and finally, after 111 ns, disappearing at diametrically opposite points.   (b) After 2 s, there is a brief uniform illumination of the entire sphere, followed by darkness.

**26.** (a) 0.32″.   (b) 0.16″.   (c) Zero.

**27.** (a) $\sqrt{c^2 + v^2}$.   (b) c.

**28.** $6.7 \times 10^{-10}$.

**29.** (b) $2.00 \times 10^{30}$ kg.

**31.** 0.57 c.

**33.** (a) $c - u$.   (b) $c + v$.   (c) $c + 2v + u$.

**34.** (a) c.   (b) c.   (c) c.


## CHAPTER 2

**4.** (a) 0.140.   (b) 0.9950.   (c) 0.999950. (d) 0.99999950.

**7.** $x' = 138$ km; $y' = 10$ km; $z' = 55$ km; $t' = -374$ μs.

**8.** (a) $x' = 0$; $t' = 2.29$ s.   (b) $x' = 6.55 \times 10^8$ m; $t' = 3.16$ s.

**9.** 0.80 μs.

**10.** 0.80 m.

**12.** (a) 0.48.   (b) $\Delta w' = 1320$ m or $\Delta t' = 4.39$ μs.

**13.** (a) 0.866c.   (b) 2.000.

**14.** (a) $2.21 \times 10^{-12}$.   (b) 5.25 days.

**15.** 1.53 cm.

**16.** 6.4 cm.

**17.** 40 mi/h.

**18.** (c) 0.938 m; 32.2°.

**19.** 250 ns.

**20.** (a) 26.3 y.   (b) 52.3 y.   (c) 3.72 y.

**21.** (b) 0.99999943c.

**22.** $1.4 \times 10^{-8}$ s.

**23.** 55 m.

**24.** 0.991c.

**25.** $4.45 \times 10^{-13}$ s.

**26.** (a) 180 m.   (b) 750 ns.   (c) 0.80c.

**27.** (a) Zero; 495 m; 1360 m; 4630 m.   (b) Zero; 396 m; 594 m; 653 m.

**28.** (a) 26 μs.   (b) The red flash.

**30.** S′ finds the flashes to be 3.46 km apart and finds that each flash occurs 5.77 μs later than the flash just beyond it.

**33.** $t'_1 = 0$; $t'_2 = -2.5$ μs.

**34.** (a) S′ must move towards S, along their common axis, at a speed of 0.48c.   (b) The "red" flash (suitably Doppler-shifted).   (c) 4.39 μs.

**35.** 2.40 μs.

**36.** (a) 4.00 μs.   (b) 2.50 μs.

**37.** $4.0 \times 10^{-13}$ s, the wavefront from BB′ arriving first.

**39.** (a) The S′-frame is seen by S to move in the positive x-direction with a speed of 0.899c.   (b) Also the "red" flash (suitably Doppler-shifted).

**42.** (a) $5.8 \times 10^5$ m².   (b) Event 1: −225 m, 4.45 μs; Event 2: 1050 m, −0.50 μs; $5.8 \times 10^5$ m².   (c) Event 1: 6869 m, 23.3 μs; Event 2: 5669 m, 18.6 μs; $5.8 \times 10^5$ m².

**43.** (a) $-1.9 \times 10^5$ m².   (b) 436 m.   (c) A frame moving in the direction of decreasing x at a speed of 0.90c.   (d) No.   (e) Spacelike.

**44.** (a) 2.5 μs, in all three frames.   (b) A frame moving in the direction of decreasing x at a speed of 0.53c.   (c) No.   (d) Timelike.

**47.** (a) 34,000 mi/h.   (b) $6.4 \times 10^{-10}$.

**49.** 0.81c.

**50.** (a) 0.81c, in the direction of increasing x.   (b) 0.26c, in the direction of increasing x. The classical predictions are 1.0c and 0.20c.

**51.** 0.95c.

**52.** 0.54c.

**53.** 1.025 μs.

**54.** (*a*) 0.35*c*.   (*b*) 0.62*c*.

**55.** Receding at 0.59*c*.

**56.** 0.88*c*.

**57.** Seven.

**59.** (*a*) 0.817*c*; along the *x*-axis.   (*b*) 0.801*c*; 3.58° forward from the *y*-axis.   (*c*) 0.801*c*; 3.58° backward from the *y'*-axis.

**61.** (*a*) $v_{AB}$ = 0.93*c*, in a direction 31° north of west.   (*b*) $v_{BA}$ = 0.93*c*, in a direction 31° east of south.

**63.** (*a*) 0.866*c*.

**64.** 23 MHz.

**66.** 0.80*c*.

**67.** 0.0067 nm.

**68.** Yellow (551 nm).

**69.** (*a*) − 28.8 nm; − 204 nm; − 393 nm.   (*b*) − 29.5 nm; − 236 nm; − 472 nm.

**70.** (*b*) 0.80*c*.

**71.** (*a*) 482.903 nm.   (*b*) 489.385 nm.   (*c*) 0.011 nm.

**73.** + 2.97 nm.

**74.** (*a*)  + 2.97 nm.   (*b*)  + 0.90 nm.   (*c*) − 2.19 nm.

**78.** (*a*) 43.9°.   (*b*) 10.2°.

**80.** 10.2° in each case.

**81.** (*a*) Zero.   (*b*) 43°.   (*c*) 87°.

**82.** (*a*) 0.067.   (*b*) 10.2°; 7.0°; 2.2°.

**83.** (*b*) 6.8°.   (*c*) 0.996*c*.

**85C.** (*c*) 6.25 m.   (*d*) 3.13 μs.   (*e*) 1.09 × $10^9$; timelike.

**87C.** (*a*)  + 157 nm.   (*b*) − 1.49 kHz.

## CHAPTER 3

**1.** (*a*) $V_0/\gamma$   (*b*) $\gamma m_0$.   (*c*) $\gamma^2 \rho_0$; 0.10*c*.

**2.** (*a*) 0.13*c*.   (*b*) 4.6 keV.   (*c*) 1.2%.

**3.** (*a*) 79.1 keV.   (*b*) 3.11 MeV.   (*c*) 10.9 MeV.

**4.** (*a*) 0.943*c*.   (*b*) 0.866*c*.

**5.** (*a*) 10.9 MeV.   (*b*) 43%.

**6.** (*a*) 1.00 keV.   (*b*) 1.07 MeV.

**7.** 1.13, 5.59, 25.1, 112, 504, 2254.

**8.** (*b*) 3.88 km/s.   (*c*) 6.3 cm/s.

**9.** (*a*) 256 kV.   (*b*) 0.746*c*.   (*c*) 1.50$m_0$.   (*d*) 256 keV.

**12.** (*a*) ~2 × $10^4$.   (*b*) 9.4 cm/s.   (*c*) 20 GeV.

**13.** (*a*) 0.42 μm/y.   (*b*) 1.8 × $10^{-16}$ kg.   (*c*) 2.8 m.

**14.** (*a*) 0.0625; 1.00196.   (*b*) 0.941; 2.96.   (*c*) 0.999 999 87; 1960.

**15.** (*a*) 0.9988; 20.6.   (*b*) 0.145; 1.01.   (*c*) 0.073; 1.0027.

**17.** (*a*) 1580 km.   (*b*) 1.2 GeV.

**20.** (*a*) 0.707*c*.   (*b*) 1.414$m_0$.   (*c*) 0.414$m_0 c^2$.

**21.** (*a*) 126 MeV.   (*b*) 69 keV.

**24.** (*a*) The photon.   (*b*) The proton.   (*c*) The proton.   (*d*) The photon.

**26.** (*b*) 5.9 GeV/*c*.   (*c*) 501 GeV/*c*.

**27.** β(= *u*/*c*)   0.80 0.90 0.99  0.999 0.9999
  *E* GeV       1.56 2.15 6.65  21.0   66.3
  *p* GeV/*c*    1.25 1.94 6.58  21.0   66.3

**28.** (*a*) 5.71 GeV; 6.65 GeV; 6.58 GeV/*c*.   (*b*) 3.11 MeV; 3.62 MeV; 3.59 MeV/*c*.

**29.** (*c*) 207$m_e$; the particle is a muon.

**30.** (*a*) 0.948*c*.   (*b*) 649$m_e$.   (*c*) 226 MeV.   (*d*) 314 MeV/*c*.

**32.**

| Proton | Frame | *K*(GeV) | *E*(GeV) | *p*(GeV/*c*) |
|--------|-------|----------|----------|--------------|
| A | S | 5.711 | 6.649 | 6.583 |
| B | S | 0 | 0.938 | 0 |
| A | S' | 0.949 | 1.887 | 1.637 |
| B | S' | 0.949 | 1.887 | − 1.637 |

**35.** 11.1 ns.

**36.** (*a*) 2.04 *u*; 0.385*c*.   (*b*) − 38.4 MeV.   (*c*) − 38.4 MeV.   (*d*)  + 38.4 MeV.

**37.** $M_0$ = 2.5$m_0$.

**38.** (*a*) 2.12$m_0$.   (*b*) 0.33*c*.

**40.** (*a*) 0.58$m_0 c$.   (*b*) 0.20$m_0 c$.   (*c*) 2.92$m_0 c^2$.   (*d*) 2.86$m_0$.   (*e*) 0.059$m_0 c^2$.

**42.** (*a*) $u_1$ = 0.220*c*; $u_2$ = 0.724*c*.   (*b*) $K_i$ = $K_f$ = 0.50$m_0 c^2$.

**43.** Sam's observations are: (*a*) ''Jim and Sue are 8 km apart.''   (*b*) ''The speed of Jim's neutrons is 0.882*c*.''   (*c*) ''I agree with Jim.''   (*d*) ''One of Jim's neutrons was scattered through an angle of 12.1°.''   (*e*) ''Jim is firing neutrons at the rate of 12,500 s$^{-1}$.''

**44.** (*a*) 1.1 × $10^8$.   (*b*) 1.5 × $10^{32}$.   (*c*) 14; 40,000.

**45.** (a) 3.53 cm.  (b) 12.0 cm.  (c) 38.2 cm.  (d) 121 cm.

**46.** 4.00 u; probably a helium-4 nucleus (alpha particle).

**47.** (a) 330 mT.  (b) 5.89.

**48.** 660 km.

**49.** ~2 × $10^{11}$ m.

**50.** $10^4 - 10^5$ ly.

**51.** (a) 9.6 × $10^{15}$ eV = 9600 TeV = 1.5 mJ. (b) 1.0 × $10^7$.

**52.** (a) 2 × $10^{14}$ eV = 200 TeV.  (b) 4 × $10^8$.

**53.** 92.1 MeF (7.68 MeV per constituent particle).

**54.** 20.6 MeV.

**56.** 139.5 MeV.

**57.** (b) 13.8 GeV.  (c) 5330 GeV.

**58.** (b) 202 GeV.  (c) 49.1 GeV.

**59.** (a) 4.43 MeV/c.  (b) 880 eV.

**61.** (b) 0.511 MeV.  (c) 938 MeV.

**62.** (a) 1.26 × $10^{13}$ J.  (b) 3.5 h.

**63.** 2.62 mg.

**64.** ~5 μg.

**65.** (a) 2.8 × $10^{14}$ J.  (b) 3.2 days.

**66.** About one part in 3 × $10^{11}$.

**67.** 88 kg.

**68.** 6.65 × $10^6$ mi, or 270 earth circumferences.

**69.** 190 tons.

**70.** 18 smu/y.

**71.** (a) 2.7 × $10^{14}$ J.  (b) 1.8 × $10^7$ kg or 18 kilotons (metric).  (c) 6.0 × $10^6$ times.

**73.** (a) 1430 eV/c; 45.3 keV/c; 2.46 MeV/c; 2.00 GeV/c.  (b) 4.52 MeV (electron); 13.3 keV (proton).  (c) 0.99882, 0.407, 0.145.  (d) 2.04 MeV, 423 MeV, 3750 MeV.

**75.** (b) 10.4 MeV, 102 MeV.

## SUPPLEMENTARY TOPIC A

**1.** (a) The present.  (b) Spacelike.  (c) None possible.  (d) 335 m.  (e) Yes; 0.67c. (f) No.

**4.** d'/d = 1.29.

**5.** (b,c) x' = 751 m; w' = −115 m (t' = −0.385 μs).

**6.** (a) The w axis.  (b) The w' axis.  (c) The w' axis.  (d) x = 0; x' = −βw'; x = βw; x' = 0.

**7.** (a) w' = −βγx.  (b) The x' axis. (c) w = βγx'.

**8.** "3.46"

**9.** 1.73 units.

**10.** (a) No.  (b) Yes; the plane itself is such a frame. (c) Timelike.

**11.** (a) −0316c.  (b) −0.143c.

**12.** (b) In S', x' = 2.02 and w' = 1.15; in S",x" = 1.67 and w" = 0.167.

## SUPPLEMENTARY TOPIC B

**3.** (a) 0.99999950c.

**5.** (a) 4.0 ly.  (b) Nine years and four months after Bob left.

**6.** (a) 5.0 y.

**7.** (a) (40/41)c.  (b) 10 y.  (c) (40/3) ly.

**8.** (a) 6.0 y, for each clock.

## SUPPLEMENTARY TOPIC C

**4.** 4.0 × $10^{-6}$ $m_e$.

**7.** 259 d; 6.2 h.

**8.** (a) 3.9 ns.  (b) 150 ns.

**9.** (b) 600.11 nm  (c) 50 km/s.

**10.** (a) 1.25 × $10^{-3}$ nm.  (b) ±3.84 × $10^{-3}$ nm.

**11.** +0.022 Hz.

**13.** (b) 9 mm; 3 km; 2 light-weeks.

**14.** (b) Earth:    2.0 × $10^{30}$ kg/m³;    Sun: 1.8 × $10^{19}$ kg/m³;  Milky Way galaxy: 8 × $10^{-4}$ kg/m³. (c) 2.7 × $10^{38}$ kg; this is about 1/1000 of the mass of the Milky Way galaxy.

**15.** 7 arc seconds.

**16.** 4.0 arc seconds/century.

**19.** (a) 243 m. (b) 40.5 m; 1.78 gee/m.

# APPENDIX

## 1. Some Physical Constants

| | | |
|---|---|---|
| Avogadro constant | $N_A$ | $6.022 \times 10^{23}$ mol$^{-1}$ |
| Bohr radius | $a_0$ | $5.292 \times 10^{-11}$ m |
| Boltzmann constant | $k$ | $1.381 \times 10^{-23}$ J/K |
| | | $8.617 \times 10^{-5}$ eV/K |
| Elementary charge | $e$ | $1.602 \times 10^{-19}$ C |
| Mass-energy equivalent | $c^2$ | $9.315 \times 10^{8}$ eV/u |
| Permeability constant | $\mu_0$ | $1.257 \times 10^{-6}$ H/m |
| Permittivity constant | $\varepsilon_0$ | $8.854 \times 10^{-12}$ F/m |
| Planck constant | $h$ | $6.626 \times 10^{-34}$ J $\cdot$ s |
| | | $4.136 \times 10^{-15}$ eV $\cdot$ s |
| Rydberg constant | $R_\infty$ | $1.097 \times 10^{7}$ m$^{-1}$ |
| Speed of light | $c$ | $2.998 \times 10^{8}$ m/s |
| Stefan-Boltzmann constant | $\sigma$ | $5.670 \times 10^{-8}$ W/m$^2 \cdot$ K$^4$ |
| Universal gas constant | $R$ | $8.314$ J/mol $\cdot$ K |
| Wien constant | $w$ | $2898$ $\mu$m $\cdot$ K |

## 2. Some Conversion Factors

Mass
  1.000 kg = 2.205 lb (mass); 453.6 g = 1.000 lb (mass)
Length
  1 m = $10^6$ $\mu$m = $10^9$ nm = $10^{10}$ Å (angstrom) = $10^{12}$ pm = $10^{15}$ fm (fermi)
  1 m = 39.37 in. = 3.280 ft; 1 in. = 2.540 cm; 1 mi = 1.609 km
1 parsec (pc) = 3.262 light-years (ly) = $3.086 \times 10^{16}$ m
  1 Astronomical unit (AU) = $1.496 \times 10^{11}$ m
Time
  1 s = $10^6$ $\mu$s = $10^9$ ns = $10^{12}$ ps
  1 d = 86,400 s; 1 (tropical) year = 365.24 d = $3.156 \times 10^7$ s
Angular measure
  1 rad = 57.30° = 0.1592 rev

Speed
   1 m/s $=$ 3.281 ft/s $=$ 2.237 mi/h; 1 mi/h $=$ 1.609 km/h
Force and pressure
   1 N $=$ $10^5$ dyne $=$ 0.2248 lb (force); 1 lb (force) $=$ 4.448 N
   1 Pa $=$ 1 N/m$^2$ $=$ $1.451 \times 10^{-4}$ lb/in$^2$ $=$ $9.872 \times 10^{-6}$ atm
   1 atm $=$ $1.013 \times 10^5$ Pa $=$ 760.0 Torr $=$ 14.70 lb/in$^2$
Energy and Power
   1 J $=$ $10^7$ erg $=$ 0.2388 cal $=$ 0.7376 ft $\cdot$ lb $=$ $2.778 \times 10^{-7}$ kW $\cdot$ h
   1 eV $=$ $1.602 \times 10^{-19}$ J
   1 horsepower (hp) $=$ 745.7 W $=$ 550.0 ft $\cdot$ lb/s
Magnetism
   1 T $=$ 1 Wb/m$^2$ $=$ $10^4$ gauss

# 3. Some Mass-Energy Conversion Factors

|  | kg | u | MeV | J |
|---|---|---|---|---|
| 1 kg $= 1$ | | $6.022 \times 10^{26}$ | $5.610 \times 10^{29}$ | $8.988 \times 10^{16}$ |
| 1 u $= 1.661 \times 10^{-27}$ | | 1 | 931.5 | $1.492 \times 10^{-10}$ |
| 1 MeV $= 1.782 \times 10^{-30}$ | | $1.074 \times 10^{-3}$ | 1 | $1.602 \times 10^{-13}$ |
| 1 J $= 1.113 \times 10^{-17}$ | | $6.700 \times 10^9$ | $6.241 \times 10^{12}$ | 1 |

$$c^2 = 931.5016 \text{ MeV/u} = 8.987552 \times 10^{16} \text{ J/kg}$$

# 4. Some Rest Masses

|  |  | kg | u | Mev/$c^2$ |
|---|---|---|---|---|
| Electron | $e$ | $9.10954 \times 10^{-31}$ | 0.000548580 | 0.511003 |
| Muon | $\mu$ | $1.88357 \times 10^{-27}$ | 0.113429 | 105.660 |
| Neutral pion | $\pi^0$ | $2.40598 \times 10^{-28}$ | 0.144889 | 134.965 |
| Pion | $\pi$ | $2.48806 \times 10^{-28}$ | 0.149832 | 139.569 |
| Proton | $p$ | $1.67265 \times 10^{-27}$ | 1.00728 | 938.279 |
| Hydrogen atom | $H^1$ | $1.67356 \times 10^{-27}$ | 1.00783 | 938.791 |
| Neutron | $n$ | $1.67495 \times 10^{-27}$ | 1.00867 | 939.573 |
| Deuterium atom | $H^2$ | $3.34455 \times 10^{-27}$ | 2.01410 | 1876.14 |
| Helium atom | $He^4$ | $6.64659 \times 10^{-27}$ | 4.00260 | 3728.43 |

$$m_\mu = 206.77 \, m_e \qquad\qquad m_\pi = 273.13 \, m_e$$
$$m_\pi^0 = 264.12 \, m_e \qquad\qquad m_p = 1836.2 \, m_e$$

# 5. Some Series Expansions

$$(y + x)^n = x^n + \frac{n}{1!} x^{n-1}y + \frac{n(n-1)}{2!} x^{n-2}y^2 + \cdots \qquad (x^2 < y^2)$$

$$(1 + x)^n = 1 + nx + \frac{n(n-1)}{2!} x^2 + \frac{n(n-1)(n-2)}{3!} x^3 \cdots \qquad (x^2 < 1)$$

$$\sin x = x - \frac{x^3}{3!} + \frac{x^5}{5!} - \cdots$$

$$\cos x = 1 - \frac{x^2}{2!} + \frac{x^4}{4!} - \cdots$$

$$e^x = 1 + x + \frac{x^2}{2!} + \frac{x^3}{3!} + \cdots$$

# Index